Terry

There is no possible way I can thank you enough for sponsoring this work with Industrial Press. Certainly you fit page 1.1.1 paragraph 3 as well as my comments in the Acknowledgements

Steve Thomas
7-20-01

Successfully Managing Change in Organizations:

A User's Guide

Stephen J. Thomas

Industrial Press Inc.

Library of Congress Cataloging-in-Publication Data
Thomas, Stephen J.
Successfully managing change in organization: a users guide/
Stephen J. Thomas.
p. cm.
Includes bibliographical references and index.
ISBN 0-8311-3149-7
1. Organizational change--Management. I. Title.
HD58.8 .T489 2001
658.4'06--dc21 2001024836

Industrial Press, Inc.
200 Madison Avenue
New York, NY 10016-4078

First Edition, August 2001

Sponsoring Editor: John Carleo
Interior Text and Cover Design: Janet Romano
Developmental Editor: Robert Weinstein

Printed in the United States of America

10 9 8 7 6 5 4 3 2 1

Dedication

First and foremost, this book is dedicated to my wife Susan. Her belief in my ability is constant and has more than helped me persevere when I didn't always believe in myself. Her love, encouragement, and support are the key success factors in my life.

This book is also dedicated to Peter, Michael, Jen, and Allison. There have been so many ways that you have supported my work, even when you didn't realize it. With a loving family believing in me, how could I fail to accomplish my dreams?

Finally, over the last 30 years I have had the opportunity of working with many fine and talented people. All of you, and you know who you are, have taught me a great many things. This book is the sum total of that knowledge. Dedicating this book to all of you is my way of saying thanks for the excellent working relationships we have enjoyed.

Acknowledgments

I would like to express my sincere thanks to the people who helped with the book: Peter, Michael, Jen, and Allison Thomas, Robert DiStefano, Bette Bernstein, and Larry More. The book is better as a result of your efforts on my behalf.

For my friend and sponsor Terry Wireman, thanks for your help. Had you not taken time to introduce me to Industrial Press, this work may never have appeared in print.

For my publisher Industrial Press Inc., and specifically John Carleo, I want to thank you for having faith in my work and the value that it will deliver. Also my thanks to Janet Romano for all of her hard work – from designing the cover to converting my many files into a well composed book.

For Robert Weinstein, thank you for helping provide clarity and focus in order to deliver a better product.

Preface

For most of my career I worked in what is commonly referred to as a reactive work environment. What that means is that I would come to work in the morning, spend a little bit of time finding out what went wrong overnight (we were a 24–hour-per-day operation), and then spend the remainder of the day running in circles in an effort to inefficiently and ineffectively correct the problems. By the end of the day I would go home extremely tired and very frustrated at my inability to work on the more strategic issues that, ultimately, would help us improve.

Then an amazing thing happened. I was assigned to a staff position somewhat removed from my former reactive world. I would come to work and actually find time to focus on strategic initiatives. Rather than spend inordinate amounts of time inefficiently chasing plant problems, I was able to apply my time to improvement initiatives. My world had changed. I had time to think and, as a result, add real value to our business.

The reason I am telling you this is that many of the things that I worked on were initiatives to move us away from the reactive to a more planned and proactive way of thinking about work. In short, I was working on change initiatives that were designed to replace my old world with something far better. Now it is entirely possible that you might not have the same opportunity. However, I will bet that the work environment that you are in could use a change for the better. It could probably stand some changes that would provide more leadership and would improve how work is conducted, how the organization is structured, and how people communicate and interact with one another.

The questions that you are faced with in a massive effort of this sort are establishing where to begin, what needs to be done, and what are some of the better ways to do it. How to devise the answers is what I am delivering in this book. I have been where you are now. I have struggled with trying to figure out what to do and how to do it. At the time I wished that I had a book like this to help me through the process. Well, I may not have had it, but in the chapters, pages, and examples that follow, it is there for you.

One of my goals in writing this book was not just to provide information but additionally to create a two-way dialog so that the book would not be a one-time event. To this end, I have built a web site that is accessible through the site maintained by Industrial Press, Inc. After you read the book and have thought about all that it contains, I would like to hear from you and open a dialog about your change efforts, both successes and failures. While this idea is not conventional, it is far from unattainable in the world of the Internet. However, there needs to be a starting point. That point begins when you buy the book and begin your work in changing what you have now into something better.

Contents

Part One
Getting Started

Welcome to a book that I sincerely believe will help you and your company make order out of the very complex process of organizational change.

If you are reading this, you obviously have an interest in the subject. Most likely you are involved with, or impacted by some form of organizational change within your company. Like most of us in business, you have probably witnessed many change efforts during your career. Some may have failed before they started; some perhaps started and then failed. Maybe even a few succeeded. Probably one or more of these shoes fit, because like it or not, change is one constant in business as well as in our individual lives.

Each company that has gone through the process of change has learned from the experience and ultimately improved, whether that particular effort succeeded or failed. Even failure can provide a useful "bad example" of how not to do it the next time, and there always is a next time. Those resisting this very basic element in the life of a company are either out of business, or are on their way out. This point also applies, on a more personal level, to individual careers.

What I have attempted to do by writing this book is to provide a "user's guide" for those of you involved in the process of change. I have been where you are now. I offer you ideas and concepts that will provide you with the necessary tools to achieve lasting success. They will help you understand your environment and how to successfully move your process forward. The material in this book does more than start you on the road; it accompanies you on your journey.

Chapter 1
Introduction to the Process of Change

1.1 The Beginning

Several years ago, I was part of a program to introduce the concept of "quality" into our plant. For a very large fee, we hired a consultant to work with us and make our "quality process" a success. Site management imposed ground rules on the project team, most of which (in retrospect) set us up for failure. The program had a kick-off date less than three months from the time it was introduced to management. With such a tight timeframe, there was inadequate time to prepare for the meetings that were an integral part of the process. To make matters worse, the meetings were to be mandatory for everyone. The official message was, "meetings are mandatory, participation is voluntary."

The consultant set up measurements to track things such as meeting attendance, quality suggestions submitted, and number of people trained. However, these were more about how the program was functioning than about the savings we were trying to achieve. Training was provided over an extended period, yet it was not started until after the program had already begun. This was confusing to the participants who were being asked to do things for which they had not been trained. A plan to make change self-sustaining over the long term was never developed. Instead the focus was on getting it started and reporting on how well it was doing. The two leaders of the effort went to extreme, sometimes even coercive, lengths to make sure the process was functioning well on all of the designated levels: meetings, measurements, and training.

A few years after its inception, both of the program leaders retired, virtually on the same day. And guess what? The process officially died the day after they were gone. If the truth were told, the process had been dead for at least a year. The plant was just going through the motions to maintain the illusion. Because the managers believed the program was still in place, they were satisfied. They never looked for the truth; the illusion was sufficient.

I was also involved in another effort that was designed to install a computer system in a multi-plant environment. Again, the management hired a consultant whose purpose was to work closely with the project team in order to make the effort a success. One of the things that the consultant told us was that the system implementation was nothing more than an enabler of change and, that to be truly successful, we needed to address the work process first, then use the system to support it.

With a clear vision of changing the work process and then supporting it with the new system, we made a detailed presentation to senior management. We were never even allowed to finish our proposal. What senior management wanted was for us to simply install the system in the shortest time-frame possible. It was their belief that the system would foster change and generate the benefits that were assumed to be associated with its use.

The implementations were completed without the work being redesigned. The result was that the users were not satisfied with the product. The system had not been designed to function within the current work process. The lack of change in the existing process made use of the new system quite difficult. Ultimately, the project designed to reduce costs and improve productivity fell far short of expectations.

Several years later, under different management, all of the sites developed processes that improved the work, enabling the system to be used in the way that it had been designed. This led to a higher level of user satisfaction and, finally, to overall success of the original effort.

Had the company chosen to change work processes first, the original project would have taken longer. But the results would have been more significant, they would have happened sooner, and they would have been far less painful to those involved.

Clearly, the examples I have just described point out that the process of change and the degree of difficulty in making the effort a success often go unrecognized. Furthermore, even when they are recognized, insufficient attention is given to the careful planning and execution of the actual work to provide for a successful outcome.

Nothing is more prevalent in industry today than change. Some of these change initiatives happen as organizations evolve, and often require little intervention. Others are more far-reaching. They involve efforts specifically designed to improve organizational functions. You probably have experienced these process design changes in whatever business you are in, maybe more times than you care to think about. These changes have been undertaken to address competition, a changing product line, productivity improvement, mergers, plant shutdowns, and the list goes on and on. What is important to recognize is that this condition we call change is probably the one constant in business today. To further complicate matters, this change not only affects our businesses, but it has a very real and personal effect on each of us. Some of these effects can be positive, some otherwise.

We get involved in the change process in many ways and for many reasons. Some of us have been asked to lead change efforts. Some have been assigned the responsibility, whereas others have openly attempted to initiate change in order to make things better. Whatever your reason for being involved, you probably have had occasions when you knew you needed to do something but could not figure out the next step. From the personal examples that I provided at the beginning of this chapter, you can see that I have been in the same place that you are now.

When you find yourself in this position, there are only a few ways to attempt to solve your problem. The first is to work with someone in your own company who has experience with the change process, either formally or through having already done the same work that you are trying to do. These individuals can help, but their perspective is usually limited to the functional areas where they have experience.

The next and most prevalent solution is to hire a consultant. There are both good and bad points about this approach. If used correctly, however, a consultant can be of value and can help you through the process. The plus side is that what you are asking them to do is their area of expertise. They usually have a great deal of experience working with firms undergoing change. The down side is that this is their business. They most likely have created a process model that they follow, a model that may or may not fit your needs. Another problem with the use of consultants is that many firms abdicate their responsibility for the effort to the consultant. This withdrawal can hurt. As good as some consultants are, they eventually have to leave you on your own. If you have abdicated all of the responsibility for the work effort to them, the work that they have done leaves when they do.

The last approach is to take ownership of the process with support from senior managers who understand the time and the complexity required to move an effort of this sort forward. Knowledgeable resources are needed for your team; these may be found within the company, or externally in the form of a consultant.

I believe in this third approach. Taking ownership is a significant step towards long lasting success. Ownership means you will design it, you will develop the details, you will roll it out to the workplace, and most importantly you will assume accountability and responsibility for its success. The prospect of such s step is frightening, but the value that you can bring to your company and yourself is immense.

To add more complexity to the issue, work process change is not a single initiative targeted at a single problem. Usually process changes span many areas of the business. They are usually linked together, so that a change in one area affects activities in others.

It is my guess that you have the skills and the knowledge to handle these problems and manage these complexities yourself, or within the group responsible for your change process. The problem is that you may not know where to start, how to proceed, what questions to ask, or how to measure whether what you did was successful or if it caused bigger problems in other areas.

Having experienced many of these situations, I have written this book so that you and others involved in the change process

- Do not have to "reinvent the wheel,"
- Can understand and even measure the interrelationships between the various change efforts,
- Will be able to figure out the next steps on your own,
- Can recognize potential problems before they happen,
- Can persuasively communicate to your management (at all levels) regarding the need for time, consistency, focus, and understanding.

1.2 Why I Wrote This Book

When I first became involved with management of organizational change, I didn't know where to start. I didn't understand the relationships between the various elements associated with this type of work, and I didn't know how to recognize potential problems. However, more than anything else, I wanted to be successful and help my company be successful. In addition to the help that I got from the consultants we hired, I spent a great deal of time looking for, buying, and reading books on the subject. I guess I was looking for a "magic pill," just one book that would help me make sense out of the chaos of change. Many books that I read helped me conclude that change was needed, but I already knew that. Many helped me understand the difficulties and potential roadblocks to success, but I was already learning that first hand, in a very painful way. The problem was that none of the books addressed the problem globally. Nor did they cover the steps needed to help me create order, to develop a game plan, and to understand from a user's perspective how to make it all work towards a successful outcome. I was frustrated, and if you have been in this position yourself, you know exactly what I am talking about.

Then I had two opportunities that changed my situation. First, I returned to school where I learned about organizational dynamics and the complexities of the change process. Second, I began to work with someone in my company who had a clear vision of where he wanted the organization to go, how to make it happen, and how to empower others to take ownership and move the process forward.

The frustration that I had experienced in my work within the world of change management, coupled with what I feel are a lack of user-oriented guidebooks on the subject, were the two driving forces that interested me in writing a book "by a user for users." However, the most important reason is that I wanted to help those of you out there who are finding it difficult to make your change process a success. I want you to have a single source of information that will help you as you proceed.

1.3 Who Can Benefit from Reading This Book?

When I started to write this book, my focus was on those whom I will refer to as "users." To me, a user is a person who has been assigned the task of developing, designing, implementing, or working with the change process. In a sense, you are using the ideas, theories, concepts, and processes developed by your management, consultants, your team, and others to successfully carry out a change process within your business. You are the users of the ideas and plans, and it is your job to "make it happen."

As I thought more and more about the book, it occurred to me that users work at many levels. Upper-level managers are users because, although they may work at a global level, they still have responsibility for a successful outcome. Middle management are all users because they work within the organization to bring a successful change process into being. Those at the "bottom" of the management organization are also users because they have to actually implement change as part of the day-to-day work effort. Further, those at the working level (usually an hourly workforce) have to take the changes that have been envisioned by others and help make them a success.

The point is that everyone is a user when it comes to the process of change. Yet not everyone understands how they can contribute to successful change within their business.

This book was written for all the users who want to understand how it all works, what role they can play, and how components interact in a very complex dynamic system. Granted that as individuals at different levels in the organization read this book, there will be different levels of understanding and ways to apply the information. However everyone will be working with the same basic understanding, and ideas will be centralized and all focused on the same topic.

1.4 Why My Book Is Different

This book is different because it's for users, but the difference goes beyond just this simple statement. As you will learn when I introduce the chapters that follow, this text is different for many reasons;

- *It is not about convincing you of the need for change.* If you are involved in the process, you are already well past the point of recognizing that there are problems and change is required.
- *It is not about getting started.* If you are reading this with a serious interest in using it to help you through the effort, you have already started.
- *It is not written at a CEO's level by a CEO,* it is not written by a partner in a consulting firm, and it is not written as an academic work. It is written by a user for users.
- *It is not written as if change were a time limited project* to be started and taken through to completion. This process has no completion, it goes on and on.
- *It views change as a complex and dynamic system within your organization.* It addresses individual elements of the process (such as leadership), but it does not treat them as stand-alone elements.

This book is for you to use throughout your process. It will provide some theory and what I feel is a lot of practical experience and useful ideas to help you as you proceed.

1.5 How To Use The Text

If possible, you should read the book from start to finish. It was designed in four parts and they follow a sequential order. The first part discusses getting started; the second addresses concepts that will contribute to your success; the third addresses the execution phase of all that you have learned; and the last discusses how to move forward after the initial effort.

However, there may be readers who have an interest or specific need to go directly to one of the later sections or individual chapters. It is not a problem if you read the text out of order. However, if you take this approach, you should realize that information contained in previous chapters may be needed if you are to fully understand some of the things I am describing in subsequent ones. To help make this easier for you, Section 1.6 describes what is included in each of the chapters. If you choose to skip around, this annotated index will help you to keep from missing important and linked parts.

1.6 What Is Included—A Look Ahead At The Chapters

This book is written in four parts, each divided into individual chapters addressing specific material relevant to that part. What follows is a brief description of the content of each chapter. You should be able to identify specific chapters of interest as well as see the overall logical sequence in which this book was assembled.

Part One: Getting Started

Part One focuses on laying the groundwork of the book. It's designed to bring everyone to a common level of understanding.

Chapter 1 Introduction to the Process of Change

This chapter provides the introduction to the entire subject by focusing on the fact that change efforts can be successful. All too often, people's experience is to the contrary. The discussion centers on what the book is about, why it was written, what makes it different from other books on this subject, and a brief look at what is included in the chapters.

Chapter 1

Chapter 2 Laying the Foundation

When people speak about work in the change arena, they believe that the terminology they use is immediately understood by every one else involved in the effort. This is not true and leads to confusion and misinterpretation. This chapter lays the groundwork so that you can proceed with a common understanding of terminology as you work your way through the book and, eventually, in your individual change efforts.

Chapter 3 Change: Not Linear, But Nonlinear

Most people, when they think of a change process, think linearly. They think of a starting point, a logical process to be followed, and then a point at which the process is complete. Change doesn't work this way. While there is an apparent starting point—the time when you begin thinking about change—the process is nonlinear, and it never ends. This chapter focuses on the concept of spiral learning. In this mode, you plan what you will do, execute it, and then reflect on the process and outcome before you take the next step. It is nonlinear because each time you execute a spiral you learn, and the next step may be something which was not even conceived when the process started.

Chapter 4 The Vision of the Future or How Do We Know Where We Are Going So We'll Know When We Have Arrived

Without a vision, there is no common understanding of where the change process is to lead the organization. This chapter addresses the need for an overall vision and who needs to develop it. It also addresses how the vision needs to be communicated, understood, and ultimately owned by those who have to "make it happen" with in the organization.

Part Two Concepts Which Will Contribute To Success

Part Two discusses many important concepts that must be under stood to prepare for and conduct the change effort.

Chapter 5 From Vision To Reality

Change efforts should not proceed without a vision of the future. Every member of the organization needs to have some under standing of where the company wants to go. The hard part is translating this vision into something tangible that employees can accomplish and, at the same time, clearly see how their efforts support the work of the company. This chapter makes that connection. It introduces the Goal Achievement Model, which uses the vision and goals set by upper management to create a framework for more concrete initiatives, activities, and measures.

Chapter 6 The Roadmap Of Change

As an organization moves along a nonlinear path to reach its vision, a roadmap is needed. Its purpose is to act as a guide along the pathway to the vision. The roadmap illustrates the organization's progress and helps avoid obstacles along the way. The roadmap also links the diverse components of the change effort so that everyone can understand the big picture.

Chapter 7 A Case For Teams

In today's businesses, no one works independently. We rely heavily on others to help us accomplish any activity or project on which we are working. Therefore, some form of teamwork is necessary for success. This chapter addresses and supports the valuable role teams play in successful change efforts. This chapter looks at the concept of critical mass; which states that you do not need to have everyone embrace change for the change effort to take hold.

Chapter 8 Working with Consultants

All too often, a company hires consultants, pays for interviews and studies, and finally receives the report—only to do nothing with the results. There is a better way. Consultants can play an important role, that adds value to the change effort. However, you need to have a clear understanding and a proper working relationship to get the results you want. This chapter discusses how to use consultants, while stressing the importance of ownership by the organization.

Chapter 9 Resistance To Change

Every change initiative will be met with resistance. Among the many reasons is the fact that people are often not comfortable with change. Yet in today's business world, change is part of the equation. To succeed with a change effort, you need to transform resistors into supporters. Otherwise, resistance will be an ongoing problem, slowing down or even destroying the change effort. This chapter addresses the problem of resistance and discusses ways to overcome it.

Chapter 10 The Web Of Change

Books that address only specific components of the change process miss the larger picture. Change has many components; they must all be equally addressed in order to achieve success. Furthermore, work done in one area affects all of the others. In fact, a positive change in one area could even have negative or disastrous effects in others. This chapter introduces the Web of Change. The eight elements of the web are then covered in detail in Chapters 11 and 12.

Part Three Execution: It's More Than Magic

Part Three moves to the execution phase of the change process. It identifies and discusses in detail the eight elements that comprise the effort.

Chapters 11 and 12 The Elements

Throughout this book, change has been viewed as an effort comprised of many integrated elements. They consist of leadership, work process, structure, group learning, technology, communication, interrelationships, and rewards. These two chapters discuss each of these so that the reader can see why they are important, how they fit into the effort, and how they fit together. A survey that is included in the appendix and on a disk helps readers to create their own web diagram.

Chapter 13 Fitting The Elements Together

Chapter 10 introduced the concept of the Web of Change. At this point, the elements of change have been discussed along with the survey that generates the data points for the web. Chapter 13 uses the web to elaborate on the interaction of these elements, analyze their interactions, and evaluate what to do with the results.

Chapter 14 How to Handle the Process Of Change

This chapter addresses the process of change itself. It addresses the following questions: How do you change? When do you change? How extensive is the change? Who should be involved in the change?

Chapter 15 Measurement

A key component of a successful change effort is measurement. The idea that "what gets measured and receives attention gets done" is explained. Most organizations measure results monthly based on their criteria and existing financial procedures. This chapter proposes that to measure a change process, monthly measures are not always the best approach. Instead weekly measurement is more useful and can have a positive impact on the change effort.

Part Four Completion And A New Beginning

This last part of the book ties everything together. It explains various methods for moving forward, both to begin the process and once the process evolves.

Chapter 16 Techniques Of Continued Growth

Many projects have a clear beginning and ending. A change effort, however, may have a beginning, but it never ends. The process needs to continually grow and improve upon itself. This chapter discusses core competencies, skills acquired from doing the work that cannot be duplicated by competition. It also looks at work process assessment and redesign, methods that help provide focus on the process.

Chapter 17 Conclusion

This chapter summarizes the text. It also describes the web site that will be used as a method to create a dialog with the readers.

1.7 Let's Get Started

I hope that you will find this guidebook useful, not just as a place to start or as a text to provide the initial information for your effort, but also as a book which will help you through all the phases of the process and into the future. I wish you success in your efforts and offer you this quote from Nicolo Machiavelli;

> "There is nothing more difficult to take in hand, more perilous to conduct, or more uncertain in its success than to take the lead in the introduction of a new order of things, because the innovator has for enemies, all of those who have done well under the old conditions, and luke-warm defenders in those who will do well under the new."

Chapter 2
Laying the Foundation

2.1 Foundation

Before we can begin to do meaningful work in the area of change management, or for that matter, in any area where numerous departments, groups, and people are involved, we must develop a common foundation. By doing this at the beginning, we can minimize misunderstanding along the way. To reach this level of common understanding, several issues must be addressed. Don't worry, we will get into change management soon enough. First we need to talk about some important ideas that will become the framework of how we are going to address the subject of change.

The first concept I want to discuss with you is that of basic assumptions. You may be surprised to learn that the assumptions made by good people have often led change processes to their demise. Once you understand how dangerous assumptions can be, you will be armed and ready to proceed. Never again will you take something that you encounter and process it without first validating it with facts. This is critical to moving through a good change management process.

Next I will talk about assumptions that you may have about your company, your organization, your work group, and even yourself. Following this discussion I will introduce terms that will be referred to throughout the book.

Chapter 2

This chapter concludes with discussions of several important points: the difference between strategic and tactical thinking, the distinction between incremental and step changes, and the dispelling of the belief in a "magic pill."

2.2 Assumptions - How They Keep Us Misinformed

A long time ago, someone told me that the word assume could be broken into three parts. The idea was that when you assume something you make an ass of u and me. At the time, I didn't understand exactly what the person was telling me. Over the years however, I have learned many times at great personal pain that this simple statement is very true. Assumptions are a major problem in the area of change management. When you are trying to change something — and almost always change has to be made through people — it is dangerous to assume anything. Invariably you will be wrong and the process may suffer as a result.

A story I once heard at a training session clarifies the problem of making a wrong assumption. A man was driving his car along a winding country road on a beautiful spring day. Because the weather was so nice, he had his window open and was admiring the view. Suddenly, over the hill in the opposite direction, came another car. The driver was leaning out of the window, obviously trying to get his attention. Upon seeing this, the man slowed down. As the other car drove past, the driver screamed, "Pig!" The man's first reaction was one of surprise, immediately followed by anger. He could not understand why this stranger would insult him in such a way. The harshness of the insult ruined his day. He hit the gas. Speeding over the hill on his way home, he ran into a very large pig standing in the highway. The wrong assumption got him into a lot of trouble.

Another more personal example took place many years ago with my first supervisory assignment. It was a hot day and I went out to check on a maintenance job in the plant. Upon arriving at the job site, I noticed one of the mechanics laying on the ground with his shirt off and his eyes closed. My initial assumption was the man was resting on company time, and getting a sun tan as well. My assumption was incorrect; he was having heat stroke. It was then that I think I learned what that person had meant long ago when he talked about the meaning of assume. It was my first lesson, though not my last, about why assumptions are dangerous.

Now that I have provided you some of my own examples, take a few minutes to think about assumptions you have made over your career. You probably have as many painful examples as I do. The point is that making assumptions about people or things can get all of us into a lot of trouble.

Armed with this insight, we can now discuss why we make assumptions and how they cause the problems they do. The model that describes the phenomenon relating to assumptions has been called the Ladder of Inference (see Figure 2-1). I first encountered it in Chris Argyris's book *Overcoming Organizational Defenses.* There is a good description of the model here, but I think an even better one is in Peter Senge's book *The Fifth Discipline Fieldbook.*

Using the image of a ladder, you start on the first rung with something you observe. You then work your way up the rungs, each time adding

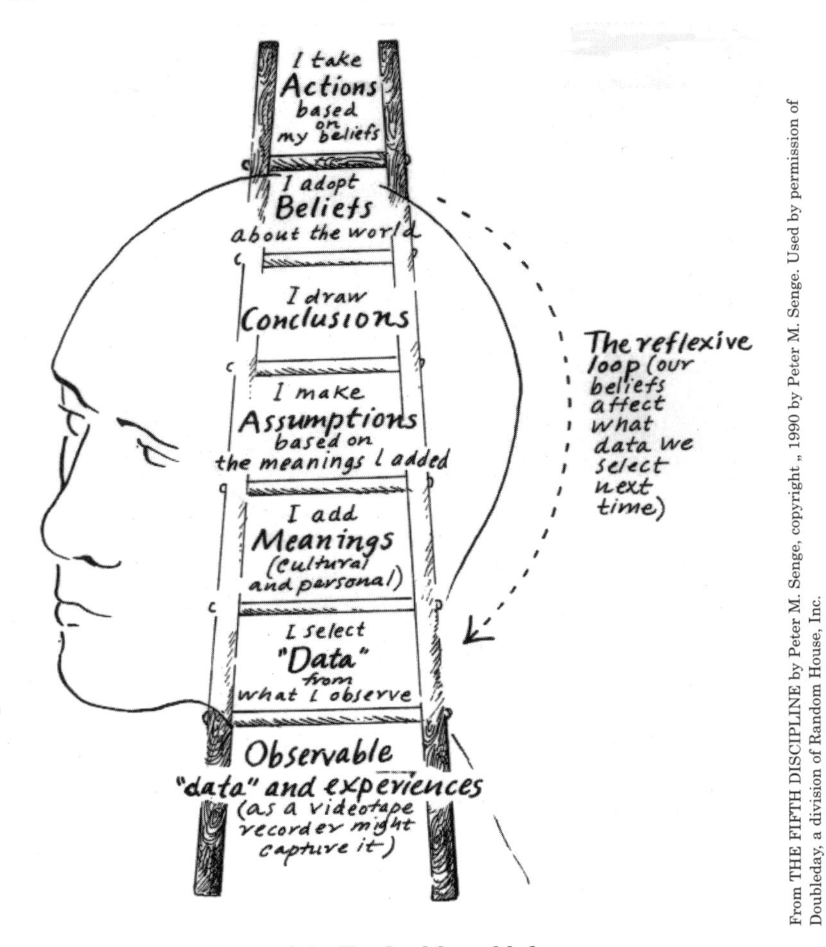

Figure 2-1 The Ladder of Inference

another layer of abstraction to the observation, until conclusions are drawn and actions are taken.

The ladder has seven rungs. On the first rung, we experience observable data and have experiences as if a videotape recorder was actually capturing what was happening for playback at a later time. On the next rung, individuals select data from what was observed. This selection is quite natural because we can not absorb all of the data we receive as input. Consequently, we need a screening or filtering process so that only those things we consider important are stored.

On the third rung, we add meaning to the data we have stored. Here is the step where we begin to introduce error. The meaning that we add is personal, individual, and — most of all — different from the meaning that anyone else would add. The reason is that the meaning we attribute to an experience is based on our cultural experiences, our individual past experiences, our learning, and especially our beliefs. As you can see, the meanings we add are diverse and based on many variables. If two people experience the same event, they will probably interpret it differently.

On the fourth rung of the ladder, we make assumptions based on the individual meanings that we added to the data. From these assumptions, we move to the fifth rung and add our conclusions. Thus far, we have taken an actual experience and, by adding the multiple levels of abstractions and filters, quite possibly distorted the experience far beyond what it originally meant.

On the sixth rung, we adopt beliefs about our world. Finally on the seventh and last rung, we act on these beliefs and create experiences for others to climb their own ladder.

Another part of this model is equally dangerous. It is called the Reflexive Loop. As you can see in Figure 2-1, the beliefs we adopt at rung six work as a filter for the data we select the next time we have a similar experience. Therefore, erroneous assumptions from a previous time can hamper our ability to select the right information the next time. The Reflexive Loop also makes incorrect assumptions harder to dispel.

What is most disconcerting is that our brain works this process with lightening speed; the only observable parts of the process are the data at the outset and the action as a result. The rest has gone on inside our heads. Quite often, because of the rapidity of the processing, we might not even be aware that it has happened.

We need to learn how to avoid making wrong assumptions. The simple answer is assume nothing! Instead, question what you see, get facts straight, and then decide based on those facts how to proceed. If you can keep off of the ladder, you can keep going in the right direction. As Sergeant Friday used to say on *Dragnet,* "please give me the facts mam, just the facts."

Let's go back to the example of my first supervisory experience and see how I used the Ladder of Inference to draw the wrong conclusion. First I showed up on the job. The initial data that I received as input was a view of the entire job site, the work, and the mechanic lying on the ground with his shirt off. All of this was observable data. The data that I selected as important was related to the mechanic. I had seen hundreds of jobs, but the mechanic at this job site stood out. In my experience to that point, no one lies around at a work site (selected data). I then added some of my own meaning to the now filtered data. The man was relaxing because that is what it looked like to me. From this meaning I made additional assumptions: Not only was he relaxing, but he was relaxing on company time and getting a good tan in the process! My conclusions, the next rung up the ladder, were that he was lazy and not producing as well as the other mechanics on the job. If I had proceeded to the next rung I could actually have adopted a belief that all mechanics who were not working were non-productive. If this conclusion had been drawn, I eventually could have acted on these beliefs and run my organization in an entirely different manner — one of distrust.

Without further validating my assumptions, every time I saw this mechanic, the Reflexive Loop would kick in and I would see him as a non-performer. I would use my data from the previous experience to prejudge his performance regardless of the circumstances. Applied on a larger scale, this belief structure could ruin an entire organization.

What can we do to stop this process from getting us into trouble? Even more important, what can we do to avoid causing irreparable damage to a very delicate process of change? The simple answer is to get the facts first, before the Ladder of Inference can take hold. We need to ask questions before we get to the step where we make assumptions. We need to take the observable data and filter it to make it manageable. At the same time, we need to stop the process from continuing until we understand the data.

Suppose I had asked the mechanic, "Why are you lying down?" His response would most likely have been "I have been working and I don't feel very well." Armed with this fact, I would not have assumed that he was sleeping on the job. Instead I would have drawn a far different conclusion: that he was sick from the heat and needed medical attention. In the first scenario, I have developed a bad opinion of a good mechanic; in the second, I have helped the man get the attention he needed. Same circumstances, different outcomes.

It should be obvious how this story relates to our subject of change. Because you will be interacting with many departments, groups and individuals, you must get the facts first. If you don't assume anything, you will have a far greater chance of success. Let us now examine some of the more common assumptions about change.

2.2.1 The Company, Organization, and the Work Group

In each of our businesses, we face many assumptions that deal with the company, its leadership team, our own organization, and our work group. Some of these are;

- They don't really care.
- They don't "walk the talk"
- They are not really interested in making things better.
- They never tell us what's going on, so we don't know.
- They expect us to change, but they won't.
- They have a new program, but if you wait long enough it will go away.
- They never listen.

This list, though abbreviated, makes a point. They sound very much like conclusions and beliefs from the top of our Ladder of Inference.

Take a minute to examine any one of these statements. What observable data have you experienced that led you to these conclusions? What filters did you place on the data? What experiences have led you to these conclusions? You may find that many of your assumptions are based on past change efforts that have failed, past bosses who really didn't care, or other events that have provided you with your own set of meanings that you attribute to the behavior of others. Now take a little more time to ask what real facts you have to support your conclusions. Have you asked clarifying questions to confirm your data before reaching conclusions about the nature of the change process?

If you are going to be a key player in your change process, you need clarity. Then you will not have doubts about the process itself. In addition, you will understand that others are at this moment drawing their own conclusions; they are moving up their own Ladders of Inference relating to the process of change in which you are involved.

It is entirely possible that some of your facts will prove that departments, groups or individuals really don't care to change. This conclusion should not be dismissed lightly. It is a very important piece of information, one that should cause you to focus even more of your effort on the problem groups. However, the conclusion should not be an assumed truth, but a truth based on hard fact.

2.2.2 Change and Its Impacts

We also make assumptions about the impact of change on the organization. Many of these assumptions, like those concerning the company, are based on people's experiences and not on known facts. Some of these assumptions include:

- People will lose their jobs with change.
- I may get demoted.
- I will be forced to do different work (and I may not like it).
- Things have been fine. Why change?
- If we change, we will ruin a good and well-functioning organization.

These assumptions, whether individual or group fears, are usually not based on factual information. While it is true, especially in today's business climate, that change sometimes does lead to reorganization, demotion, and layoff, most companies do not take these matters lightly. Even though there is negative impact on individuals, companies often are forced to make these changes to survive. In my working life, I have lived through and survived numerous reorganizations and layoffs, as well as three sales of the company for which I worked. I have been part of the group making these assumptions and have found that, invariably, the results were never as bad as I had assumed that they would be. Working in the world of erroneous assumption caused a great deal of lost time, excessive worry, and wasted effort.

The main point is that you want to have as much information on the table as soon as possible in order to give people facts. These facts will stop people from jumping onto the Ladder of Inference, drawing wrong conclusions, and then making matters worse by acting on these conclusions.

2.2.3 The Level of Commitment

Assumptions are made about the level of commitment required to make a change process a success. Many assume that if we work very hard for a short time we can make the change, then sit back and bask in the light of our accomplishment. This assumption is wrong. Change is an ongoing process. It doesn't end and you can never simply sit back and rest on your laurels. This is why change is continually referred to as a process.

Quite often people fail to understand the level of commitment needed to make the process a success. They believe that being involved is enough. Again this assumption is wrong. The difference between commitment and involvement is best described by a breakfast of bacon and eggs. In this scenario, the chicken is involved. The pig is truly committed.

2.2.4 Timing

Timing is another area where incorrect assumptions get in the way. The majority of us have what I will call a "project mentality." Our experience tells us that everything has a beginning and an end. We apply this meaning to the process of change. Furthermore, we are impatient and anxious to make things happen. These beliefs that the effort is finite and that, to be successful, we need to get it done quickly interferes with a successful change process. Change takes time.

Industry benchmarks indicate that you probably need three-to-five years to create a viable change from scratch. Patience is the key if long-lasting change is to be successful. Those who assume otherwise set in place expectations that can not be achieved. The way to overcome this assumption is to present facts supporting the concept that "slow is fast." This is especially true if the individuals you are addressing are the company's leaders. They, like everyone else, need to understand that change takes time and that time needs to be provided for success to be accomplished

2.2.5 What You Are Getting Into

The last assumption that needs to be addressed is the one about what you are getting into when you embark on the change process. You might assume that this type of effort is like a project: there is a start, a working process, and ultimately a finish. *This is not so.* What you are getting into is a very complex process which goes on and on. Just when you think you have overcome one hurdle, another even more formidable one will make itself known.

One reason is that change is invariably about altering the way that people work. Recall the Machiavelli quote at the end of Chapter 1. People do not like change. At best you are going to get lukewarm support from those who may do well and, often, outright resistance from those who feel threatened. Therefore, what you are getting yourself into is not, as some assume, a simple work project. Instead, you are getting yourself into a very complex and difficult process that will tax your stamina, your patience, and your commitment to seeing the process through, even when things seem to be falling down around you. At the same time, a tremendous feeling of accomplishment awaits if you and your organization succeed because you will have had a major hand in changing the company for the better.

2.3 Definition of Terms

Now that you can see how assumptions can hamper the change process unless they are converted into actual facts it is time to introduce a few terms. This book is about *change management,* but what does this term mean? For our purposes it means exactly what it says. It is the supervision of a process designed to alter and improve the functioning of your business. Business and the associated work processes are always changing and improving. If you are not, your competition certainly is. Without continuous improvement, they will eventually win. That's good for them, but not for you.

Another term, the Web of Change will be covered in detail in Chapter 10. The Web of Change is a radar diagram that relates the elements of change together in a unique fashion. The basic concept is that a specific set of elements exists that, if all incorporated correctly, will provide you with a totally successful change effort. Each of these eight elements, listed below will individually contribute to your success. But I especially want to show their combined total effect on the work process system. In developing this concept, the diagram of the elements recalled a spider web, hence the name Web of Change.

The eight elements of this web are:

1 Leadership
2 Work Process
3 Structure
4 Group Learning
5 Technology
6 Communication
7 Interrelationships
8 Rewards

These are the eight key areas where you need to focus your work efforts if you are going to be successful, not only for the change effort, but also in the marketplace.

Chapter 2

2.4 The Difference Between Strategy and Tactics

The next part of the foundation is distinguishing between strategic and tactical approaches to work and, more specifically, to work process change. Strategic work is work for the future. It is the development of activities that are designed to create a plan that, if followed, will lead you and your organization to an identified end point. Tactical work is the actual doing of the activities identified by the strategy. Tactics are what gets things done.

There are two types of thinking and, equally, two types of work going on every day in business. The first, strategic, is associated with the development of plans for the future and how we can prepare the groundwork now so that we can get there later. The other, tactical, is the execution of those plans to comply with the strategy.

You may be asking, "So what? We already have groups doing both strategic and tactical tasks as part of their jobs. Everything seems to be working just fine." If that's the case, then you are indeed fortunate. Nevertheless, I would suggest you look deeper into your process for possible problems. It is not usually realistic to expect someone to operate both strategically and tactically at the same time.

But why not? Why can't we have someone create strategy and then have the same person implement it on an ongoing basis? The answer is that these two types of thinking—strategic and tactical—each take place each in a different hemisphere of the brain. Strategy is the result of creative thinking, an activity associated with the brain's right hemisphere. Tactics, on the other hand are associated with the brain's left hemisphere. Well-established research concludes that the two hemispheres do not communicate with each other very well.

If you ever have concentrated on tactical activities, you will know what I mean when I say that you can't do both. If you have ever had a job that is all tactical, and then have been expected to do strategic thinking, you have probably found the transition difficult. Strategic and tactical work feel, and actually are, very different.

Years ago I was in charge of a large organization that repaired equipment in the plant. In addition to supervising daily work activities, my staff and I were given the task of creating a strategy for improving the overall work process. The effort was far more difficult to complete than expected because the group found it hard to detach from the day-to-day work. The staff was composed of tactical, left-brain thinkers who were very good at what they did. However, they found that developing a long-term strategy—a right brain task—required that they work in unfamiliar territory. Help was required to make the change. Employing a consultant versed in strategic development solved the problem. This individual who provided guidance and direction, helped them think more strategically.

In general, for a change effort to be successful you need to have separate groups handling the strategic and tactical aspects of the change. Because this is seldom a practical solution, you need to provide time and support for people to learn to think differently. In addition, you should not expect a single group to do both activities at the same time.

2.5 Incremental and Step Changes

Another part of the foundation is distinguishing between incremental and step changes. These are very different approaches to change.

Incremental change relies on individual small changes over time that eventually lead you to your original goal. The problem is that each of the steps takes time. Each runs the risk of failing or, even worse, being diverted into an area which will slow the change effort or ruin its original intent.

Step change advocates getting all of the plans made, then making the change to a totally new system or process all at once. This type of change has pitfalls as well. If the step is not well orchestrated, the result can be chaos.

Suppose your company wanted to install a new computer system that would have a major impact on all departments in your company. An incremental change approach would involve a rollout of the system, possibly on a department–by-department basis. For a period of time, both computer systems (new and old) would run in parallel. In the step change approach, the old system would be shutdown and the new started up in one step. There would be no rollout. Everyone would get the new system at once. To make a step change possible, you would need a great deal of advance planning. Either of these approaches will work. However, as a leader of the change effort, you need to fully analyze both approaches before you select one.

As a compromise, I offer a third approach. There are some processes that need a step change to get the effort started, but the adjustments can then be handled as incremental changes. The step change starts the process moving, greatly shortening the dangers and problems of implementation. However, the incremental portion takes learning into account. It allows for adjustment in small steps. When you are thinking about the process change, do not focus on the extremes. Instead consider a successful integration of the two concepts.

2.6 The Magic Pill

The process of change is difficult. It takes time, requires commitment, and is ongoing even after we have accomplished our initial goals. However, there is a belief by many that some one or some group has a "magic pill," the answer that, when applied, will instantaneously solve all problems. Many times we look for it in the individuals we assign to the change effort, the managers of our businesses or departments, or the consultants we hire. Ultimately, when the magic pill is not found, we search for someone else that may have found it. This search explains why managers sometimes go to look at successful firms that have achieved what you are trying to achieve, and why some companies spend a great deal of money for consultants. They are looking for a quick way to resolve their issues and problems.

The truth is that this magic pill doesn't exist in the form in which we are looking. It is not a one-time cure-all, but rather one of those medications you have to take over the long term if you want to get well. Our long term equates to years, but if you follow the course of treatment, there is hope.

2.7 How Do We Know What Success Looks Like?

Another question which may be on your mind as we close out this chapter is, "How do we know what success looks like?" This is a question asked by many who are entering into the very complex and difficult area of change and managing the change process.

It would be easy for me to say that you measure yourself against your competition, but that would be a wrong and simplistic answer. The truth is that you need to look at yourself and your progress within the framework of a continuously evolving vision that you have created. If you have succeeded in achieving that vision, then you know what success looks like.

We are now ready to move to a group of chapters that address vision and what we need to do to turn vision into reality. Reading these chapters will provide you with a deeper understanding of how to define success for yourself and your company.

Chapter 3
Change: Not Linear, But Nonlinear

3.1 The Idea of Nonlinear Change

Suppose that you want to build something, for example a house. Assuming that you could act as the general contractor, what would you do? First you would develop a concept for your house, followed by a design that includes engineering drawings and specifications. You would then hold a bid meeting that eventually would enable you to award a construction contract, start actual construction, accept the work and eventually move-in. This example illustrates a linear process: the work effort to build the house is performed in a sequence that, if represented by a line, would be straight from start to finish. The process has a definite beginning, when you decide you want to build the house, and a definite end, when the house has been completed and you move in. In reality, the process of change is not so straightforward. Instead, it is actually nonlinear; it doesn't proceed with one predefined step immediately followed by the next. Nonlinear change is different. Although this concept can be uncomfortable, it will help you reach your change objectives, even though you may not know the exact path when you begin the journey.

Chapter 3

3.2 What is Linear Change?

Let's start with a brief discussion of linear change. Once we have a good understanding of linear change, we will have an easier time in grasping the concept of nonlinear change.

Linear change is the process by which events take place in a known and predictable order. If you laid these events end to end, they would form essentially a straight line. The linear process is also characterized by a definite start point and a definite end point. The final element is that you can accurately predict the steps required to get you from the starting to the end point.

Linear processes are predictable throughout their entire life. Each step you take is known in advance. The results of each step, if completed correctly, will yield a result that is expected. Of course, you may have to deviate for some unforeseen problem, but the track from start to finish doesn't really change. The deviation that you had to make carries with it a subset of known steps; the linearity is perpetuated. For example, midway through building a house, you may decide to add a garage. This event was not known at the outset. Once included in the process, however, it brings with it the predictable set of steps that launches a subset of linear processes.

Another aspect of linear processes is that they are not new. They are based on past experience and essentially replicate that experience. There may be some new aspect of the process that enters into the picture, but it will usually be minor in nature. It will not affect the steps of the process or the predefined outcome.

We can define linear change as follows:

> A process with a finite beginning and ending, a path from start to finish that can be predefined, and steps that usually take place in a sequential manner so that, upon completion of one, the next is both obvious and required in order to proceed to others.

Now let's apply the definition of linearity to the process of change in an organization. On the one hand, change has a starting point. It begins when you or someone else in your company becomes dissatisfied with the current state of events and decides that a change is necessary. It is not clear, however, that the process of change, once initiated, has an ending. Unlike our house example, you can not know at the outset of a process of change exactly what the end will look like. While you may have a vision of an end state, there is no certainty that your process will look or act exactly as you predicted when you reach it.

Why is that? The environment in which you work is changing. Building a house is conducted in a static environment — one that doesn't change as you move from design through construction. This is not the case with organizational change. As you move through the process of changing the way the organization works, the environment is affected. You impact the organization with the changes you make. Furthermore the process generates learning within the organization; it can not work as it had prior to the change. Therefore, the very nature of change alters the environment. What you conceived when you started may no longer apply.

As you reach your conceptualized end state, it is highly likely that you will not really reach the end, but rather another beginning. Once again, the concept of linearity fails to properly fit the change process.

Then there is the element of the linear process that addresses prior experience. It states that the steps of a process are based on past experience, allowing you to know what is coming as you end one step and embark on the next. In the world of change, this is usually not the case. Instead what usually happens is that you take one step towards your goal and, in doing so, you generate new ideas or efforts that were not part of your original conception. However, when you take a close look at these new ideas and efforts, they appear to fit perfectly. Steps enter into the process that were not part of the original plan yet are now needed to support it.

3.3 An Alternative Defined: Nonlinear Change

We have just tried to fit the process of change into the definition of linearity and it doesn't fit on several levels. What we need instead is a concept that is referred to as nonlinear change.

Nonlinear change is an evolutionary process. It has a finite starting point, but no clear end. Actions taken at each step are generated by prior steps, affect subsequent steps, and were most likely not known or fully realized prior to the start of the step being addressed.

This concept is difficult for people who are anchored in the linear world. It asks us to believe that, as the process evolves, steps will become apparent based on our learning and the evolution of the system in which we are working. People who want to see the end point and know exactly how they are going to get there before they start may not be comfortable with nonlinear change. It asks that we take a leap of faith as we change our business environment.

Chapter 3

As with linear change, the process of nonlinear change has to have a finite starting point. This point is the place in time when you recognize that the process you have in place is not working, either at all or in the way you want. This point is the only similarity that linear and nonlinear change have with each other.

The process has no end because change is evolutionary and dynamic. Consequently, it needs to be continually growing. The alternative is to not grow and become static. Think for a minute of all of the times you have heard "It has always worked this way, so why bother changing?" Although there may be different companies with this attitude, they have one thing in common: they are probably out of business and, if not, soon will be. The road of progress is paved with the remains of companies that refused to become dynamic organizations.

Let's look at the middle of the nonlinear process. It is made up of an undetermined number of spirals. Within each spiral are four distinct parts. For definition purposes, a spiral is one cycle on the diagram shown in Figure 3-1.

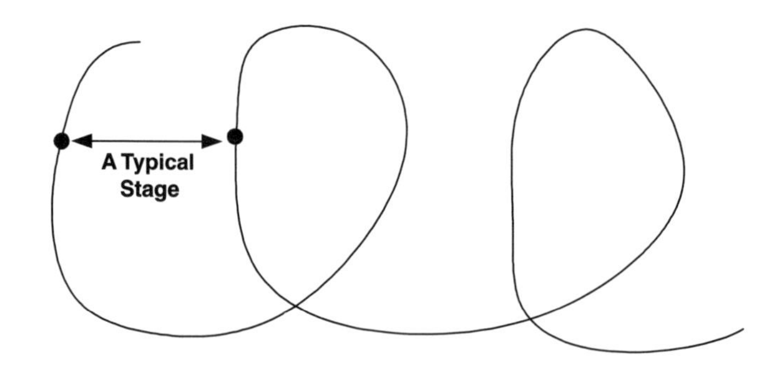

Figure 3-1 A Learning Spiral

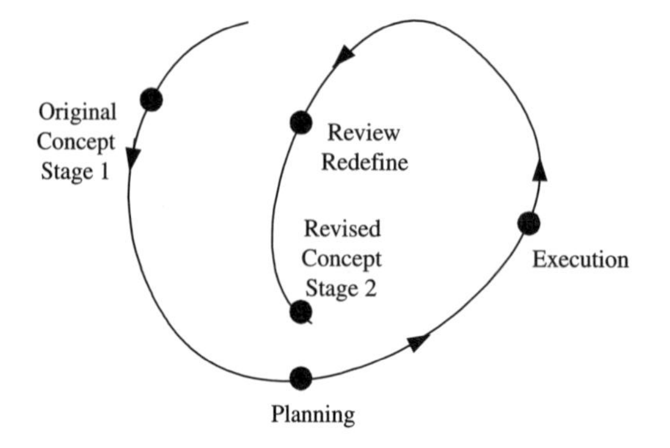

Figure 3-2 Detailed Learning Spiral Process

Chapter 3

Following the original concept, each of these spirals has four component parts: Planning, Execution, Review/ Redefine, and finally the Revised Concept (see Figure 3-2). Brief definitions of these terms are provided below.

Original Concept	This is the original idea that was developed for the change initiative. It is usually based on dissat isfaction with the current state and a desire to move to a new and improved state.
Planning	Once the original concept has been finalized, the first step is to plan how you will execute the work design. Without a good plan; you are doomed.
Execution	The next step is to actually work through the process of change designed in the original concept.
Review/ Redefine	Once the execution of the spiral has been completed, it is time to review what took place. Then redefine your original concept into something that still addresses the desired end, taking into account the changes made in the previous spiral and what you have learned.
Revised Concept	The outcome of the review/redesign step is a revised concept for the new spiral. These steps repe at themselves through all spirals in the process.

The step of reviewing what you have done and then redefining what you will do in the next spiral is generally unfamiliar to you if you work in the linear world. That is not to say it isn't a part of some aspects of your life. But it probably is not a large part of how you currently do business. If you aren't convinced, try telling your boss that you have a vision of a project's end state, but won't know the details of later spirals until the work on earlier ones has been completed.

3.4 What Are the Benefits? What Are the Difficulties?

We can now look at the benefits associated with nonlinear change. We also need to look at the difficulties from the perspective of those who have not experienced nonlinear change in the work environment. You need to know about these negatives so that you can address and overcome them in your company.

Chapter 3

First the positives. I will list them in bullet form so that they are easier to review and digest. All too often when a list is presented in paragraph form some of the value of what is trying to be said can get lost.

The benefits of nonlinear change include:

- **Allowing for and promoting flexibility.** Because you only need to know the desired end point, you and your work group have flexibility to try new things within whatever framework has been established by management. Some have characterized this approach as working within the "banks of the river." Management can broaden or narrow the distance between the banks; this process allows you latitude as long as you reach the goal.
- **Promoting group learning and allowing you to do something with what you learned.** As you progress through the spirals, your group and individual learning takes place. The iterative nature of the process allows you to learn from what you did and then to apply it to the next spiral.
- **Providing for an optimal solution to whatever change you are putting into place.** The fact that one spiral builds upon the previous one enables this optimization to take place. Compare the value of being able to make mid-course corrections to the process (the nonlinear model) with developing a plan, then rigidly sticking to it (the linear model).
- **Enabling you to minimize your chance of failure.** The ability to make mid-course corrections is a far better process. If you experience a bad spiral (the outcome was not what you wanted), you have more than enough time to make adjustments it in the next spiral.

If the above list are the benefits of using the nonlinear model and you can easily see the value, why are there still negatives? I don't believe that there are any. However, there will be colleagues at work whose eyes glaze over when you mention the nonlinear change process. Their failure to understand this concept is the negative aspect of nonlinear change. You will need positive responses to show them how nonlinear change will yield great results.

- **The process requires a leap of faith that some may not be ready to take.** While it is true that initially a leap of faith may be required, the leap is not very great. We can only improve when we recognize that the steps toward the solution continually emerge as the organization evolves through the change process. We have the vision of the outcome and the ability to participate in the review/redesign step at the end of each spiral. Therefore we maintain control over the end result.
- **Senior management loses control in working a change process in this manner.** This is not the case at all. Those involved daily with the change process have a lot more control. Being part of the creation of the vision, setting "the banks of the river," and participating in the redesign between spirals provides considerable control. In addition, the management personnel who participate actively in nonlinear change are actually more involved in the redesign than they would have been if change were introduced linearly.
- **We don't have time to operate in a nonlinear mode.** If you and your company want to survive in today's world then you don't have the time not to work in this manner. The question that needs to be asked and answered is simple: how much do you want to excel? If you really want to be the best, you need to find the time.
- **We don't have resources for this type of effort.** You need to free up resources to make change happen. The job is not part time for whoever is going to lead the process. Nor is the job for anyone other than a recognized leader who can get things done through others. Furthermore, the process needs a lot of other resource involvement throughout so that in the end everyone owns the result.

3.5 Examples of Nonlinear Change in Action

Two examples will help illustrate nonlinear change. The first is very simple and not business related: the game of chess. The second example is a business with a plant vision of high equipment reliability.

The game of chess fits most of our description of nonlinear change, except like all games it has an ending. In other ways, however, chess fits our model perfectly. Each move you make is another spiral. Every spiral is dependent upon the previous spirals and has major impact on those that follow.

You plan your move, execute it, and then reflect and redesign your next move based on what you learn from the dynamic environment of the game. The reason that it is dynamic is that your move causes a response from your opponent, changing the configuration of the board and the game. When you plan and execute your move in the first spiral, you do not definitely know the response your opponent will make. Consequently, after your opponent moves you need to reflect and redesign your next move based on a new set of conditions. This is nonlinear change! The plan you put together and the move you execute are made with an educated leap of faith that they will improve your position, and ultimately you will win the game.

The next example addresses work within a continuous process plant. In this type of work environment, equipment reliability is critical because shutdowns or slowdowns have a significant financial impact on the company. The plant's original goal was to develop a reliability program that would help identify plant assets that had a high likelihood of failure or equipment with serious production consequences. It was believed that about ten percent of the 40,000 plant assets would have real problems and fit this category. By addressing the 10%, the plant managers thought they would be able to attack and improve reliability for the critical few. Because the site had a team structure in place, the effort was designed to help the teams focus on the problem equipment and improve reliability.

This was not going to be an easy task, but it was something that was needed. The employees leading the process did not develop a detailed linear plan. Instead, with the support of the site's leadership, they approached the problem in a nonlinear fashion. Figure 3-3 describes the process they used over three spirals.

In contrast to the nonlinear approach used in this example, let's look at what might have happened using the conventional linear model. The process would have been designed and then implemented throughout all of the units and areas within the plant. If the time schedule had been tight, the original design may have had flaws not discovered until too late. Even if flaws were identified, changes to the strategy may not have been implemented. The results would have been less than desired and, quite possibly, nothing further would ever have been done with them. This may be extreme, but think about some of the work efforts with which you have been involved. Do any of them fit this description? Nonlinear change efforts can correct the problem of conducting work in this type of linear fashion.

Spiral 1

Planning	Developed a strategy for the effort, contracted for consultant sup port, and identified an area to conduct a pilotprogram.
Execution	Conducted the pilot program in two areas in the plant
Review/Redefine	Based on the execution of the pilot, the process was redesigned to improve upon what was learned.
Revised Concept	A new process emerged designed to make the effort easier to con duct and focused on obtaining the needed data

Spiral 2

Planning	Using the redesigned approach, developed a plan for the entire plant
Execution	Conducted the work effort for each unit and area within the plant site
Review/Redefine	The effort was successful. Interest level was raised in the teams and many spontaneously started working on these problems.
Revised Concept	It became apparent that the lists of problem equipment that emergedfrom the work effort supported 1. An improved work order process. 2. A more in-depth focus on critical work to be performed during plant outages. 3. A technique to focus the teams on improved reliability.

Spiral 3

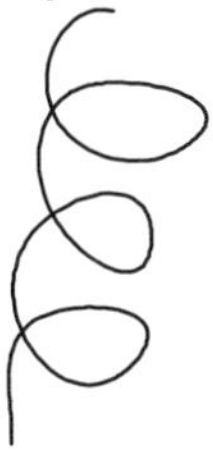

Planning	Planning in all three areas identified in the Revised Concept section in Spiral 2.
Execution	Work in progress. The results will create the next spiral.

Figure 3-3 An Example of Spiral Learning

The outcome was far different when the work was conducted using spiral learning. The original design flaws were discovered with enough time to make the corrections before the process was rolled out to the entire plant. At the end of the second spiral, it became evident that some teams were working on reliability, based on the outcome of the studies. The plant was able to capitalize on this interest. It began to implement real reliability efforts. Other spin-off benefits emerged and these were incorporated into future spirals.

When you begin a change process, do not assume that you know the answers or even what the outcome is going to look like when you start. Yes, you have a vision of the future state. However, if you proceed in a nonlinear fashion, not only will you realize your goals, but also the end result will far exceed anything that you ever expected. The trick is to provide management with an understanding of the nonlinear process. Then get their support so you can proceed. That "leap of faith" which I talked about earlier, take it! You won't be disappointed.

Chapter 4
The Vision of the Future
or
How Do We Know Where We Are Going So We Know When We Have Arrived

4.1 What is a Vision? How Far Out Should We Look?

In business as well as in our personal lives, we need a vision of some unachieved future state towards which we can strive. Without a vision of the future, there is no growth and no evolution. With no growth comes stagnation. Because we are dealing with change and the management of change, an early discussion about vision is important.

In business, if you don't have a vision, you are missing your future. Failure to create a vision for your company means that you are willing to accept the possibility of no growth; you see the current "as is" condition as an acceptable future state. Most people would not hold this belief. Most people in business recognize in today's rapidly changing world that without growth, you are in trouble.

Furthermore individuals in the company need to strive toward some goal. Without a vision, they have nothing to stimulate their individual and collective thinking in a way that will not only help them help the business grow but also help them to grow as individuals. Just as a stagnated business with no vision will soon be history, the people who are the key component of the business will be history too. When people fail to grow they become frustrated. Eventually they leave. These are the people you most need most to help move the business forward.

Consider the case of a man who worked in a plant within the maintenance organization. The plant ran very unreliably. The maintenance workforce was under pressure to repair equipment that was continually broken, restricting production.

The employees worked days, nights, and weekends just to keep the plant running. Most of the time they were successful and production didn't suffer. Every so often, however, they were not successful. Production suffered and panic ensued. To say that they were a reactive workforce would be a gross understatement of the facts. Their idea of an ideal future state was simply to be able to fix things faster, and once in a while, to be able to stop and catch their collective breath.

The manager's idea of a good day was one during which his organization was able to apply resources to all of the failures. At least on these days, he could tell the plant manager that all problems were being covered. To reach this level of success, the department spent money on more spare parts, hired additional mechanics, and, when needed, brought in expensive contractor support. Expenses were so over the budget that the manager always took vacation when it came time to report his department's expenses.

The department had no real vision of a better future where equipment failure would be an exception, not the rule. It could not see a plant where reliable equipment created an environment of continuous production and maximized profits. If they had a vision of this sort, the path they would have chosen would have been very different from the one they were following.

The result of having no vision was that the department was never able to improve. Eventually a second company purchased the company that owned the plant. When the new owners looked at the unreliable plant, they saw no option other than to shut it down. Thus the employees suffered from the same lack of vision.

This example illustrates what can happen when you don't have a vision of the future. Here conditions were so bad that the plant shut down. This doesn't mean that without a vision the same fate awaits you. However without a vision, both you and your company will stagnate.

What, then, is a vision? You can define it as follows:

> Vision is an idealized picture of a future state, one that is integrated into the organization's culture. It provides a stretch, yet it is achievable over an extended time period with a great deal of work and collective focus by the entire organization. Because it continually evolves, it is never fully achieved.

Let's look more closely at this definition.

An Idealized Picture of a Future State

Vision should be something that people can see. When pieces of it are completed, people in the company can say "what we set out to create for ourselves is what we have achieved." A picture is something that people can hold up and use to measure their progress.

Integrated into the Organization's Culture

The vision must be difficult, if not impossible, to alter so that personnel changes can not easily destroy what the collective members have created. Too often the vision is not sufficiently integrated into the company culture. In these cases, change in management can easily alter or destroy what everyone has worked to achieve.

A Stretch, Yet Achievable

The vision needs to be something that the firm can achieve, but not easily. If it is too difficult people will become frustrated and give up. If it is not a stretch, then it will be easily accomplished and will not have significant value for the group.

Extended Time and a Great Deal of Work

A vision is not something that can be accomplished in a short period of time. It represents a major shift in how a firm does business. If a vision can be achieved overnight, then it is not sufficiently a stretch for the firm. However, a worthwhile vision that takes a lot of time and requires a major shift in the culture will take a great deal of work.

Collective Focus by the Entire Organization
It is not enough if just one person, or even a few people, understand and are working to achieve the vision. Instead, the vision must be a collective effort for the entire organization. Only then can it not only succeed, but also be long lasting and beneficial.

Continually Evolving, Never Accomplished
As you learned in Chapter 3, the process is continually evolving, not according to a set plan, but instead to the process of learning. Although the vision is set at the beginning, the organization continually evolves and the end-state is never accomplished. In fact, by the time that the initially described end state is reached, a new and evolved end state will have replaced the original.

You may next be asking, how far into the future we should look to set our vision? I would answer from three to five years, depending on the company's maturity. Less than three years is too short of a time frame for meaningful change. Some firms may need two years just to get organized and begin the process of moving forward. More than five years is too long for an initial look. As you move toward the vision, you will learn both individually and collectively a great deal about process change. In turn, your vision will evolve. It quite possibly may take on forms that you might not even have imagined when you started. Therefore, five years out is about the maximum. Beyond that, your vision will change as you grow and no longer be valid.

4.2 How Visions Are Achieved

It is not the purpose of this book to take you through the process of creating a vision for your firm. Numerous books on the market already address that part of the process specifically. If you haven't yet created a vision of your future state, I urge you to obtain one of these texts and do so.

If the vision is not created properly, it will not be understood by the firm nor will it be embraced as an acceptable end state. Furthermore, the vision will not receive the attention that is required to make it succeed. As a result, it will ultimately fail. It seems then that the methods for achieving the vision are dependent upon those employed to create it. If the vision is created by a small group working in a vacuum, that group will probably have to make it work. This approach, which leads to a vision that lacks ownership, may have limited success. However, if the vision is created by teams or other large group processes, then the buy-in has a far greater likelihood of occurring; the entire group is invested in its success.

Even in the best of circumstances, achieving a vision is not a simple task. It is vulnerable to many pitfalls. Any strategy used to create the vision must translate into a strategy used to implement it. Thus the following aspects need to be considered at both stages for a chance at success:

- A strategy must be in place to take the vision from a picture of a desired future state to something that is real.
- The organization must be ready to change.
- The organization must be aligned for change.
- The vision must be extremely clear and understood by all.
- The managers, those who set strategic direction, must "walk the talk".
- All levels of the organization must make an ongoing commitment.
- Sufficient and realistic time must be allowed for the effort to be a success.

4.3 Goal Achievement: A New Way to Reach the Vision

If you put five senior managers in a room and asked them to describe how they could achieve a company's vision, you would get five different answers. The truth is that there are many methods to accomplish this very difficult task. What I will initially describe here, and in much more detail in Chapter 5, is the Goal Achievement Model.

Any vision of the future needs to be supported by goals on which the organization can work. Each of these goals, when accomplished, take us one step closer to the vision we have set for ourselves. The problem in many firms is that goal setting is just an exercise. Speaking for myself, I have set personal goals numerous times as well as worked with others to set organizational goals. Some I knew would be achieved simply doing the work. Others were stretch goals. Some of these were what I believed made sense; others may have been what my managers wanted to see.

Too often, the process of goal setting resulted in a document that immediately went into my desk drawer, only to be resurrected when I had to relate my accomplishments for the year. At that time, I pulled out the document that had been prepared twelve months earlier and desperately tried to figure out what we had done to support the goals that had been set. Sometimes many had been achieved, and I was able to breathe a sigh of relief. In other years, I was not quite as successful.

The problem with this approach to goal setting is that it is nothing more than an exercise. It is not part of the real work. Thus, goals languish in a drawer for twelve months at a time without ever really contributing to organizational progress.

If we are truly to move the organization from vision to reality, we need a process that ties together the vision, the goals set to support the vision, and the process of day-to-day work in the company. This is the Goal Achievement Model described more completely in Chapter 5.

4.4 Organizational Readiness

The clearest vision and the best planned change processes will fail unless the organization is ready to change. How do you know when an organization is ready? Sometimes the answer is obvious. For example, you probably know stories about companies on the verge of going out of business that change themselves into new and profitable enterprises. These companies were ready for change because the alternative was to shutdown and dismiss all of the employees. This example is a simple one.

How do you measure readiness, however, when you are not on the verge of going out of business? What if your firm is marginally profitable or even successful? How do you get people ready and willing to change when they don't believe in their hearts that change is needed? The tricky part is for you, who can see the future needs of your business, to get people invested in a new vision, especially when the process takes time to complete.

Three elements are needed for creating a positive climate and organizational readiness for change. These are:

- Dissatisfaction with the current state
- A vision of a new future
- A set of steps that will enable movement toward the future

These three items are linked. The first and most important item is dissatisfaction with the current state. Without dissatisfaction the seeds of change will not take root. Dissatisfaction is created by developing a good vision of the future, one that cannot be achieved in the current environment using the current processes. This gap between what is desired and what actually exists leads to the need for a set of specific steps that will close the gap.

These points may seem simple, but they are not. In the end, without some form of dissatisfaction with the current state, there will be no desire for change.

Chapter 4

4.5 Organizational Alignment

Another element that is essential to turning the vision into reality is organizational alignment. The organization must be structured in a way that will support the effort to achieve the vision. Otherwise, some aspects of the organizational structure will hamper or even work against the organization's movement to a new, usually very different, future.

As an example, let's take a company that wants to implement changes that would best be supported by a team approach. If the company is organized in a top-down style similar to the military then it could have a very difficult time assembling effective teams. Those who prefer the old structure may not cooperate, refusing to allow their employees to work across departmental boundaries. In the end, this organizational alignment will block a team process. The right organizational alignment is essential to moving change forward.

The good news is that the solution to the alignment problem is also a major step in the change process. You need to reorganize the company into a structure that supports the vision. In one fell swoop; you accomplish many goals that support the change process.

- You set up an organization that supports the vision.
- You send a clear message to the work force that things are going to be different.
- You balance advantages of step and incremental change, making a step change in the alignment that also serves as an incremental change to the total process.
- You move employees en masse to jobs where you believe they can best support the effort.

Obviously what I am talking about is reorganization. This word often brings fear into the hearts of the strongest managers. Remember, though, you are not reorganizing simply for the sake of reorganizing. (Unfortunately, all too often reorganization occurs in the vain hope that reshuffling the deck will create a miracle.) Here, this is not the case. You are reorganizing to align the company with the vision. This point needs to be clearly emphasized. Because the process of reorganization is itself such a traumatic experience, you do not want to lose sight of the overriding reason for changing the structure. Be sure your motives for reorganization are clear and are clearly communicated.

4.6 Communication: Selling the Vision

The process of communication is twofold. First, everyone in the organization needs to know what is taking place. Change is one of the things that people fear most, not just in business, but also in their daily lives. What we are proposing affects both. However, if they understand the reason for the change, their fears will be somewhat alleviated.

Second, everyone must have the same picture of the end state if there is to be any hope for long-term success. Without a shared picture, the end result will not be that which was desired. If you and I are painting a picture, and I am thinking about a landscape while you are thinking about a seascape, the end result will not look the same. This division is what we are trying to avoid.

The key to accomplishing both of these goals is ongoing communication. You need to invest the time to communicate effectively. Only then will everyone be fully informed, understand where the company is in the process, and share the same picture of what their world will look like when they get there.

Ongoing communication, however, is only the initial answer to addressing these goals. You also need training. If your firm is changing, and you truly want everyone to understand, then you will need to provide training in the new concepts. Simply informing is not enough. You must educate as well.

4.7 Walk the Talk or Lose the Group

There is a quote I remember from my youth that I never liked: "Don't do what I do, do what I say." It was one of those statements made by my parents and others when they wanted me to do something that, quite obviously, they had no intention of doing themselves. It may have been about something as simple as going to bed early. The specific example is not nearly as important as the statement itself. We would probably not make that particular statement to our workforce, but assume for a minute that we did. What do you think would be the result? How do you think the work force would view our actions? Would the organization be willing to make the effort for successful change?

The organization would probably have little respect for what we wanted. They would probably not be willing to go along with any significant change effort if they were the only ones expected to change. People view us more by our actions than the words that fall from our lips.

Consider the following "don't do what I do, do what I say" examples

- A company is in financial difficulty and cost cutting is announced, while the mangers travel first class.

- No raises are given to employees, but the managers still receive large bonuses.
- A change effort is implemented, but the leadership refuses to change.
- A reorganization takes place but certain people are protected, even if they are not the best people for the new jobs.
- A new work process is developed on paper, while the old one remains in effect.

You can probably relate to some of these statements. In fact, you can probably add many of your own. The point is that if you are a change leader, then you need to "walk the talk." You need to act in the same manner as you expect everyone else to act in the new work environment as you and your firm pursue the vision.

4.8 On-Going Commitment from All Levels Over Time

All too often, creating a vision, setting goals, and other long-term concepts get people's commitment for the moment. Then something else comes up and the longer-term nature of the work is forgotten. As you shall see in Chapter 5, the vision is the heart of the change effort. It needs to be sustained over time. Only then can the desired future be created. This long-term commitment is difficult, but essential.

4.9 You Can't Be Everything for Everyone

Earlier I wrote about the need to stay focused. As you move forward and begin to see success from your efforts, others will want you to add new items to the things that you are already doing. They may or may not be connected. If you continually add new items, your plate will eventually be so full that you will accomplish nothing. Therefore focus is extremely important. You simply cannot be "everything for everyone." It doesn't work. Overloading the process with additional work initiatives bogs everything else down. It could even destroy good work already in progress.

There is, however, a pitfall in taking a hard stance relative to additional scope. While some of the suggestions may clearly be additions, others may not. They may actually sharpen your focus and lead to a much improved work product. The trick is to be able to differentiate one from the other. This can be accomplished by closely listening to the new ideas, asking probing questions, and deciding if they should be incorporated. This approach also supports the concept of spiral learning; as you move through the nonlinear spirals new things are learned that require inclusion.

Because the entire organization is engaged in the change process, these new ideas can come equally from all levels of management as well as the hourly work force. They may have different perspectives; senior management may be more strategic, while lower levels of the organization may be more tactical. All have their own special value.

Those suggestions that do not fit should not automatically be discarded. A suggestion that doesn't seem to have a place in the change process right now may be appropriate at a later date. Such items should be placed in a "parking lot" where they can be held for the future. By storing them, you say to others that you believe their ideas are of value. However, you are not going to interrupt the current process to address them now.

This approach will work if everyone, especially senior management, understands the process for evaluating and incorporating new ideas. All too often, a suggestion from the top is treated as a directive. This response has the potential to introduce counter-productive ideas into the process. Managers need to be extremely sensitive to this potential problem, making certain that the context in which they present their ideas is clearly understood.

4.10 The Journey Begins With a Single Step

When you think about the range of issues we have just discussed, you may be overwhelmed by the enormity of the undertaking. It is true that what you are about to do is a large and very difficult effort. Yet it is not impossible. Many have gone where you are about to go; many have succeeded.

Years ago, before I got older and my knees gave out, I used to run in 10,000-meter races. I wasn't fast and never had any expectation of winning. My vision was finishing in less than one hour. When I first started doing this, I had a problem. Standing at the start and thinking about running 6.2 miles was overwhelming. How was I going to make this effort and successfully achieve my vision? The answer was found in the old proverb, "every journey begins with a single step." In my case that involved training. It also involved changing and redefining my view of the race. I no longer thought about running 10,000 meters. Instead I thought about running one step at a time. I focused my energy on placing one foot after the other in the best way I could. To my astonishment, I succeeded. You can too.

Part Two

Concepts Which Will Contribute To Success

In Part Two we are going to discuss what you need to think and plan for if your change initiative is going to be successful. You don't simply decide you want to change an aspect of how work gets completed and then have that change magically happen before your eyes. Change requires a lot of hard work. Understanding the concepts and topics addressed in this section of the book will give you a better idea of how to proceed.

Chapter 5 introduces a process that will enable you and your organization to work constructively with the vision you created. It will help you implement a process that will develop goals, initiatives, and activities to support the vision. Chapter 6 expands on this topic by addressing the outcomes and impacts of your actions, not just as they affect you, but also as they affect others.

Chapters 7, 8 and 9 discuss additional aspects of change that are important to your success. Teams, addressed in Chapter 7, are essential simply because, in today's environment, nothing can be accomplished in a vacuum. You also need to think about how or if you are going to use consultants to support your work. Chapter 8 discusses consultants in detail, showing you that, when used properly, consultants can add a great deal. Chapter 9 then discusses resistance. While change would be easier to implement without resistance, that is never the case. Chapter 10 introduces the Web of Change, a measurement technique that helps you visualize change in the context of eight elements.

Chapter 5
From Vision to Reality

5.1 The Vision and the Reality Gap

Chapter 4 looked at how vision can be created in a company. Every company at some time has probably worked on a vision statement. In fact many of you have probably worked on more than one. When completed, each of these visions was communicated to the employees with the expectation that they would turn it into reality. Most of the time, these visions fail miserably to significantly influence the company's performance. What then are the successful companies doing that others are not? How can we use what they are doing correctly to make our own efforts have real value for the company and our customer base?

The successful firms take the vision and turn it into reality. They examine their vision in light of their current reality, determine the gaps that exist, and find ways to close the gaps. They change conditions within their own firms. When completed, their vision is no longer an abstract picture, but rather the way they conduct business.

5.2 How Do We Bridge the Gap?

Most companies and individuals create goals on a yearly basis. We create them at corporate, plant, and other organizational levels. The purpose behind this yearly task is really to determine what we must do to turn a vision into reality. Goals are designed to provide the framework for closing the gap between vision and our current work situation. Many problems, however, can prevent us from closing these gaps, including:

- **Lack of understanding**
  If the vision is not clearly articulated, those trying to use it to set goals are confused regarding the desired outcome. They lose their commitment simply because they don't understand what is needed.
- **Lack of focus**
  For goal setting to be successful, everyone—top to bottom—must be focused on the effort. Without sustained focus the process will fail.
- **Lack of coordination**
  Goal setting without coordination can be very dangerous. Two or more groups working on the same goal without coordination can produce different or, even worse, conflicting outcomes.
- **Lack of a process to integrate goals**
  Goal setting will only be successful if there is a process in place to integrate it into the work process. Even at its simplest level, goal setting is difficult to achieve, and without integration, it is next to impossible.
- **Lack of ongoing feedback**
  Feedback is necessary so that groups get either positive or negative reinforcement for their efforts. They need to know if they are moving in the right direction. With positive feedback they can continue. If the feedback is negative, they can make mid course corrections to get back on track. Few things are more demoralizing than completing something you felt was a good effort and being told "No that's not what I wanted."

How do we close the gap? How do we create a process that, with management support, will ensure that goals can be successfully achieved and the vision realized? The answer is goal achievement and the use of the Goal Achievement Model.

5.3 The Goal Achievement Model

Goal achievement is a method that links organization, department, team, and individual goals, initiatives and activities to the overall vision in a way that all can easily see how their efforts contribute to the end result. This also provides a global view — a way to see how efforts by one group can affect, or be affected by the efforts of others.

Goal achievement overcomes many of the problems discussed in Section 5.2. It provides a clear understanding of the vision; for without the vision, goals cannot be properly set. Goal achievement requires focus as well as coordination among departments and groups. By the very nature of this coordination, goal achievement provides feedback to the participants.

The Goal Achievement Model (Figure 5-1) illustrates the method. Before looking specifically at how the model works, let's first clearly define its key components: the mission, goals, initiatives, activities, and measures.

Mission. The mission is a single broad statement that describes the overall vision for the company, plant, or department. The mission states the vision in a way that employees can readily understand its relevance to them. It is aimed at a high-level purpose. For example, the mission might be "to operate the facility in a reliable manner so that the requirements of the customers are always satisfied."

Goals. Goals are broad-based statements that support the mission in long-term, but specific ways. Typically, there will be several goals in support of the mission with each one addressing a different aspect. Continuing with the mission described above, the goals could include:

1. Develop a comprehensive reliability program.
2. Improve the level of workforce skills.
3. Train the workforce to make decisions focused on reliability of the equipment and the processes.

These goals actually help to sharpen the mission, focusing on the word *reliable.*

Initiatives. Initiatives are statements describing long-term efforts that will be made to accomplish a specific goal. Typically each goal will have several initiatives associated. Many different efforts are usually needed to accomplish a stated goal. Initiatives are generated by the groups who do the actual work. Including them in a visual model allows the various groups to see what others are doing. The model also provides a mechanism to avoid duplicated work or efforts that are counter productive. For the goal of developing a comprehensive reliability program, specific initia-

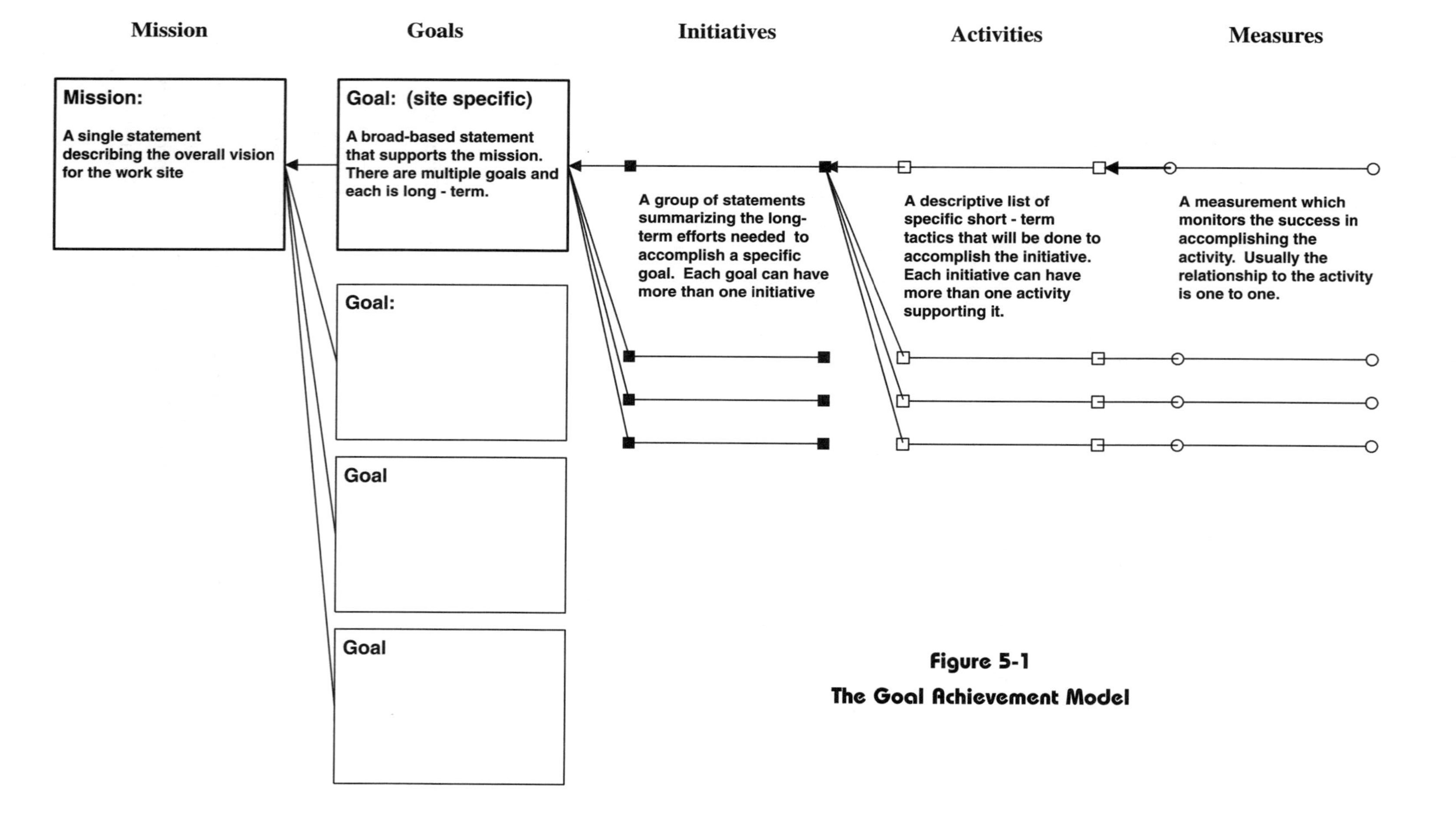

Figure 5-1
The Goal Achievement Model

tives could include:

1. Establishing a program for predictive maintenance.
2. Establishing a program for preventive maintenance.
3. Developing a tracking tool so that work can be scheduled.

Notice the increasingly specific nature of these statements as we go from mission to goals to initiatives.

Activities. Activities are the short-term, specific tactics that explain exactly and in detail what the group or individuals will do to accomplish each initiative. At this level, specific work steps and tasks are described. The activities are usually developed by the group responsible for carrying them out. Looking at the initiative of establishing a program for preventive maintenance, activities could include:

1. Determining which types of equipment should receive preventive maintenance.
2. Identifying the equipment and gathering data to load into the database.
3. Determining how to staff this activity.
4. Establishing how often each piece of equipment should receive preventive maintenance.
5. Developing a schedule for the work.
6. Developing a plan to monitor the completion of all of the tasks and to enforce the maintenance schedule.

As you can see, activities are short-term specific steps that you take to complete the initiative. Responsibility can be assigned to an individual or group for completion of the task. Without assigned responsibility, the effort gets lost in the everyday work.

Measures. Measurement is the last, but perhaps the most vital component of the process. It tracks progress for each initiative and activity. Measures show if the work is on track and hold everyone accountable. We will discuss measurement further in Chapter 15.

5.4 How the Goal Achievement Model Works

The Goal Achievement Model allows for a mission and high level goals to be clearly stated and then converted from an abstract concept into something that can be clearly understood. This conversion is achieved through the continual development of the mission (or vision) into successive levels of detail through the model. The mission of "operating the facility in a reliable manner" is abstract; it has various meanings for different groups and people within the company. If we did not proceed further into the model, leaving people to create goals on this statement alone, the result would probably not be what was wanted. However, as you look at the subsequent levels and examples, you can see that the approach becomes very specific. The abstract nature disappears.

Each successive level becomes more detailed so that by the time you reach the Activity level, the work to be done is easily translated into actual work tasks. The model enables you to identify specific tasks that, when completed, support the mission. At the lower levels of the model this is clearly achieved

Because the work done at the Activity level is recognizable in its contribution to the overall mission, employees can see the value of their efforts. Figure 5-2 illustrates how each level of the model supports the next

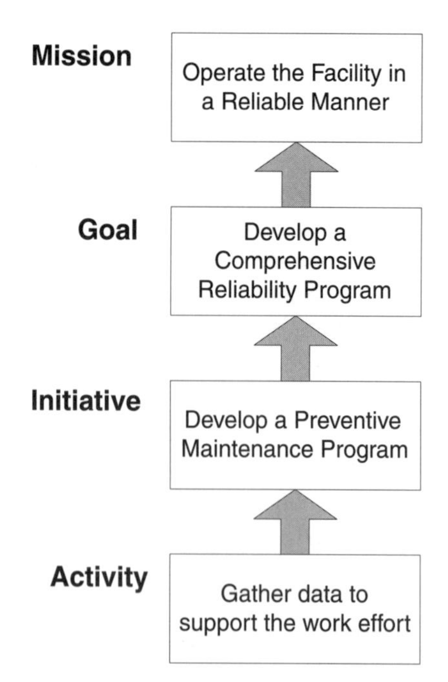

Figure 5-2 How Activities Support the Mission

higher level. For example, the activity of gathering the data supports the initiative of developing the preventive maintenance program. In turn, this initiative supports the goal of creating the comprehensive reliability program that ultimately supports the overall mission. This connection is clear not only to the executive level but also to all who are involved. Thus, the model can help drive the success of the process.

Because groups within the facility can see what others are doing, they can eliminate conflicting activities or even those activities that negatively impact each other. The majority of companies today do not have an overabundance of resources. Therefore, you want to be sure that you and your group are working on the right things. When everyone's efforts are shown, the model allows each group to focus on the work that adds value for the company.

The measures established to track the activities provide evidence that progress is actually being made. How many times have you prepared goals, only to have them wind up in a desk drawer for future retrieval when you are asked, "How are you doing with your goals?" You find yourself scrambling to see what you have done, trying to make your accomplishments fit what you said you were going to do. Proper measures avoid this problem. If you report your measures on a regular basis, the chances that you will be scrambling at the end of the year are minimal.

5.5 Goal Achievement: An Example

Let's now consider a full scale example, one that focuses on an area of importance for all companies, especially those with production facilities. Suppose you are working in a plant that does not have a good safety record. Management decides it wants the plant to focus on improving its safety record, including the overall safety of the personnel and assets. The Goal Achievement Model is put into action. Therefore, plant management develops a mission for the site that states, "Provide a safe work environment." For a site that does not already have a good safety record, this mission is impossible to achieve in a short time. Using the Goal Achievement Model over time, however, will enable the organization to focus on all the components that will help it to succeed.

The mission can have many associated goals. This mission has several, identified in Figure 5-3. The first, which will be further developed in this example, is to create an education program that raises the level of safety awareness in the plant. A second goal, to identify and correct unsafe acts,

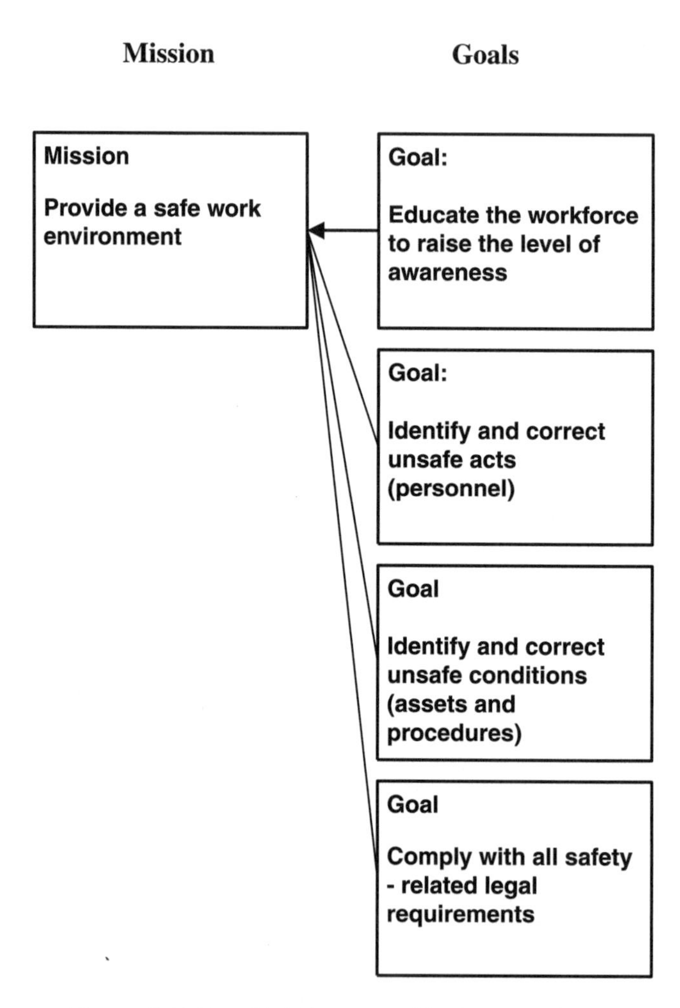

Figure 5-3 The Link Between Mission and Goals

focuses on employee actions whereas a third goal, to identify and correct unsafe conditions focuses on equipment and procedures. The last goal, to comply with safety-related legal requirements, looks at the plant in a broader environment. Note that the goals are still somewhat abstract, yet they have begun to sharpen the focus on the type of work that will be needed.

The next step, setting the initiatives, gets more specific, though it retains a long-term perspective. These steps should be done at the various organizational levels in the plant. Those who have to carry out initiatives and convert them into activities should have some ownership over the process. If the staff sets the initiatives, then it will be difficult for the employees at lower levels in the organization to feel that ownership (The exception to this rule is for any initiative that may be specifically required at the corporate level. In this case, the staff sets the initiative and mid-level management must make it successful.)

In our example, the goals shown in Figure 5-3 were presented to the plant personnel in order to increase their understanding and support. They formed teams to identify the initiatives needed to achieve these goals. Figure 5-4 shows the initiatives developed for the first goal to "educate the workforce to raise the level of awareness."

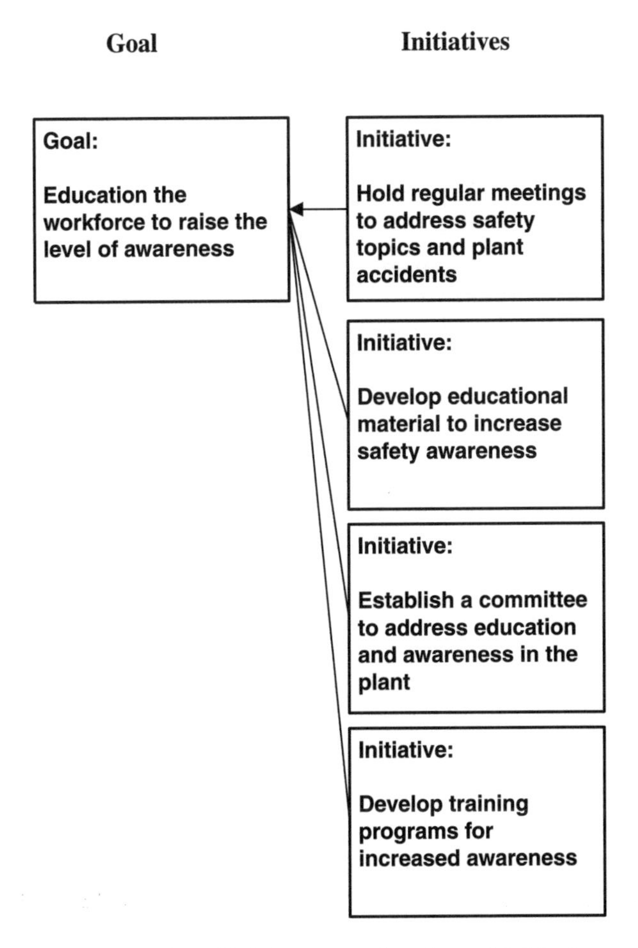

Figure 5-4 The Link Between Goals and Initiative

These are examples of initiatives within the goal of creating safety awareness. One of the better ways of generating this list is through group brainstorming, and then ranking the list by priority. Note that this goal has several initiatives, as would the other goals.

With initiatives established, the next step involves determining what activities (tactics) need to be developed. Figure 5-5 zooms in on the activities identified to support the initiative: "Hold regular safety meetings to address safety topics and plant accidents."

The trick is not just to generate this list of activities, but also to establish a set of measures that track their progress. To complete the example, I selected the activity: "Determine frequency dates, and times for the meetings." Figure 5-6 shows these measures that go with the activity.

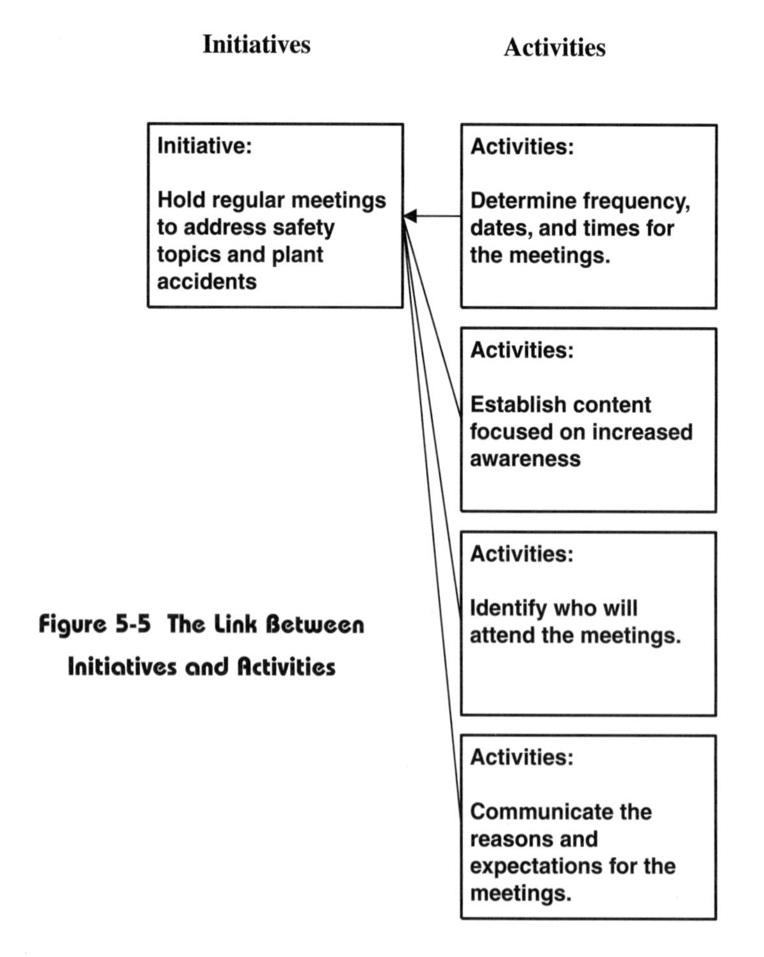

Figure 5-5 The Link Between Initiatives and Activities

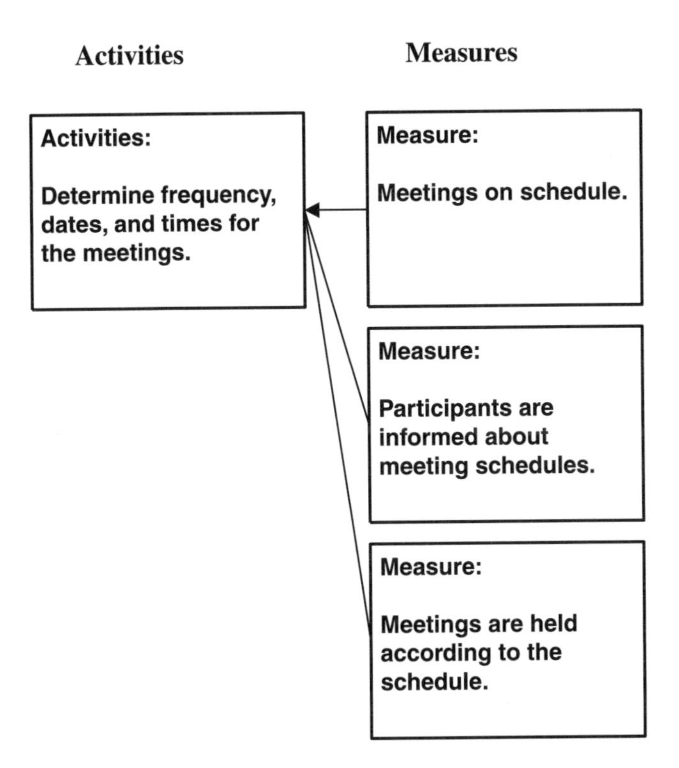

Figure 5-6 The Link Between Activities and Measures

To complete this example, Figure 5-7 provides a full diagram of all of the steps we have discussed. Take some time and see how the various steps flow from the original mission. Note how the accomplishment of a single activity can be tracked back so that those doing the work can clearly see how they are contributing.

5.6 You Think Goals Are Just a Yearly Thing

Chapter 3 introduced the process of spiral learning—how you create an idea, plan and execute it, reflect on it, and use the outcome to reframe your plan for the next spiral. This process also applies to goal setting and the Goal Achievement Model.

The vision and associated mission that we develop usually cannot be changed in a year. Consider the safety example. Do you think that a plant with a horrible safety record can change that quickly? Probably not.

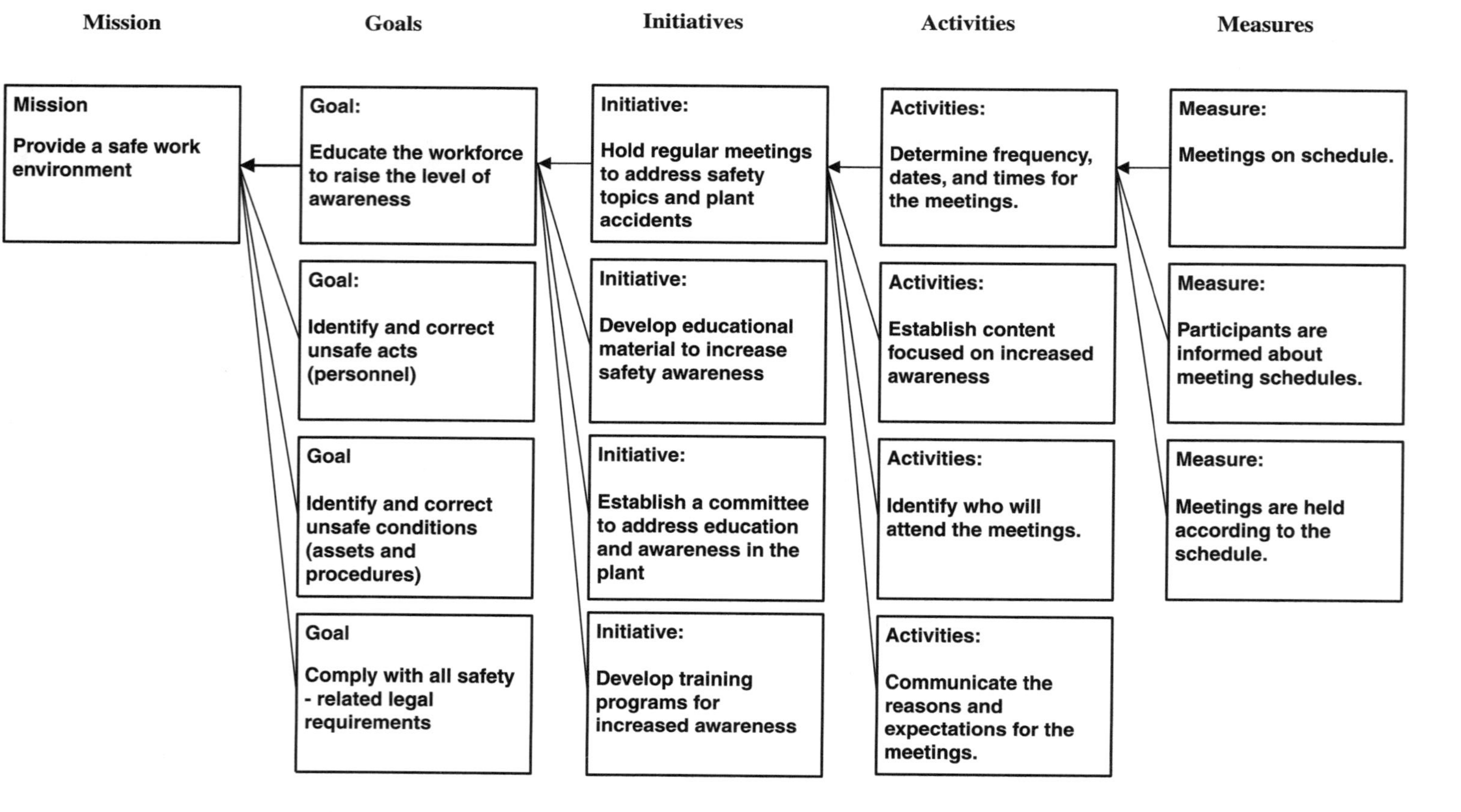

Figure 5-7 The Plant Safety Example Using the Goal Achievement Model

Chapter 5

What do we do to try to accomplish a difficult task of this nature? We set goals, initiatives, and the shorter-term activities that can be measured in smaller time frames. We have already discussed some of the obstacles that can get in the way of this effort.

From here we need to use what we learned in Chapter 3 and in this chapter. We start with the model as described. We then breakdown the high-level vision and mission into smaller units that can be completed by the site personnel. We also recognize that accomplishing the vision requires using the spirals from Chapter 3. At the end of the year do more than look at your goals. Reflect on what you have achieved. Use what you have learned to reframe your goals, getting one more step closer to the vision in the coming year. The goals and all of the associated components are not just a one-time event. They are one of many spirals to get you where you want to go.

Chapter 6
The Roadmap of Change

6.1 Why a Roadmap?

The purpose of a real roadmap is to show you how to get from where you are to where you want to be—safely and with minimal problems. The Roadmap of Change has the same purpose, but its focus is helping you to traverse the pathways of the change process. As you turn your vision into reality, this roadmap enables you to take a hard look at what you are doing, in your own area of influence as well as others. Like a roadmap in your car, the Roadmap of Change can help you avoid hazards along the way.

It also serves other purposes. It provides a way for you to communicate what you need the organization to do so that the change initiative will be successful. You cannot simply paint a visionary picture. You also need to show people specifically how to move from where they are to where they want to be. The roadmap shows them in the form of a work plan.

The Roadmap can provide the organization with a unity of purpose. It serves as a way to measure the progress you are making. Like any map, it has milestones that can be identified and used to measure progress. The milestones provide reinforcement that the workforce is making progress along their planned route.

Chapter 6

The Roadmap can also help groups working on similar goals to interact and focus on a common direction. Two or more groups may be working toward the same goal, but without interaction they will usually produce more than one result. In this case, each group may be satisfied with its achievement, but you, as the manager, now have to address the different results. This is a real problem. How do you blend the answers these groups have developed without having them lose ownership or, worse, becoming dissatisfied with the process? The answer is to never let this problem occur in the first place. You avoid it with the Roadmap, which leads everyone to the same place.

However, sometimes groups working toward the same goal end up with *conflicting* results. Correcting this type of problem uses up valuable energy of the people involved. An additional problem in resolving such conflicts is that people who could add value and who are essential to the change process could buy out of the process as a result of having their solution rejected.

The Goal Achievement Model allows organizations to break a vision into goals, initiatives, and activities. For organizations with many departments, however, the model by itself doesn't provide a mechanism to keep the initiatives and activities from overlapping or conflicting. The Roadmap provides this mechanism.

Suppose you are in charge of maintenance in a plant that uses an assembly line process to manufacture goods. In any plant of this type, the maintenance group has a very difficult task; the line must be operating to enable it to meet the demand for its products. Suppose further that this plant is reactive in its approach to problems. In other words, things break, the line shuts down, and the maintenance organization attacks the problems in order to restore the line to service. This kind of reactive maintenance has neither planning nor proper job preparation. All it has is the knee-jerk reaction of the workforce to whatever is broken. Because this approach is non-productive and inefficient, the organization decides to shift to a proactive maintenance approach. In the new work environment, the focus will be on preventing failure. If failure does occur, work will be well planned before it is executed. I can tell you from experience that this transition is not an easy one to make. It is, however, one that companies are trying to accomplish every day.

The company's vision, then, is to establish a new way of conducting maintenance. The mission might be to "prevent line failures by increasing equipment reliability of the equipment." If we follow the process described in the last two chapters, we can use the Goal Achievement Model to develop specific goals, initiatives, and activities. Figure 6-1 shows one model that might be developed, following the third goal to the initiative level.

Let's take a look at the initiatives we have created. If we simply develop activities for each initiative, we could be setting ourselves up for failure. How is that possible? Other departments in the plant, without understanding and buying into your goals and initiatives, may be developing their own. Even though these may still support the vision, they could conflict with what you are trying to create.

Mission	1. Prevent line failures by increasing the reliability of the equipment.
Goals	1. Establish a program to predict failures before they occur. 2. Establish a preventive maintenance program to eliminate unplanned failure. 3. All work is to be planned with reliability of the equipment as the primary focus.
Initiatives (based on goal #3 only)	1. Plan all work prior to execution. 2. Create a planning group for development of work plans. 3. Execute work per the plans.

Figure 6-1 Goal Achievement Model: Maintenance Group

Let's examine this idea by considering the first initiative in Figure 6-1; "plan all work prior to execution." Activities that fit this initiative could include: establish planning standards (what makes a good plan), set up planning templates for repetitive work, and requiring job planning prior to work being sent to the field.

But now consider the Operations Group. It is working with same mission, but has independently created different goals and initiatives, as seen in Figure 6-2.

We have conflicting goals across the departments. Operation's The two Goal Achievement Models starts with the same mission, but it immediately diverges. Each is directing the organization in a different way. For example, Maintenance would like all work planned before execution to optimize the use of their limited resources. Operations, however, wants Maintenance to be responsive to line failures. This divergence can create a

Mission	1. Prevent line failures by increasing the reliability of the equipment.
Goals	1. Establish and program operators to monitor equipment closely. 2. Have maintenance mechanics responsive to line failures.
Initiatives (based on goal # 2only)	1. Assign maintenance mechanics to the line to make timely repairs. 2. Have mechanics work under the direction of Operations.

Figure 6-2 Goal Achievement Model: Operations Group

serious problem when you want a change for the better, not a change for the worse. Here is where the Roadmap can provide the solution, sorting out the differences and providing a single direction.

6.2 How Does a Roadmap Fit into the Change Process?

With this example in mind, the Roadmap is the tool to align change efforts within the organization, to eliminate conflicting goals, and to keep the change process on track. It is the third stage of a process that begins with establishing the vision, developing higher level details with the Goal Achievement Model, and maintaining focus and clarity with the Roadmap. A successful change effort can not succeed without all three of these pieces being properly put into place and correctly used.

The vision is the starting point. If the different groups are not aligned at this level, then alignment throughout the rest of the process will not be possible. Therefore the vision must be clearly articulated. Everyone must understand what it means. Furthermore, spot checks should be made on a continuous basis to ensure alignment with the big picture. An easy way to make these checks is to have an oversight team, usually senior management, maintain the validation process. This team should evaluate the goals of all the groups participating in the change process. If these goals are in alignment across departments and groups, then everyone probably understands the vision in the same way.

The second step is to ensure alignment of the goals from the participating groups. Although the goals may not be precisely the same, the oversight group can make certain that they all support the plant vision. Those who have developed the goals should discuss them with the oversight team and other participating groups. A large presentation format is often the approach to this type of effort. Even though many people don't like to sit through these presentations, this format makes certain that the oversight team, as well as others who have set goals, initiatives, and activities understand the process, not just in their own areas, but in others as well. This understanding takes them beyond their own boundaries to a more global awareness of what the organization is trying to achieve and how their component fits into the bigger picture.

At this time, you should also make sure that the goals of each group support those of the other groups at the site. A good set of independent goals lose their value if they disable goals of other groups.

The large group format must go beyond simply presentations. Where conflicts exist, they must be addressed before they are turned into initiatives and activities. Once the process has moved beyond the vision and goal level, it becomes much harder to undo. It is better to attend to conflicts sooner rather than later.

Next we need to make sure that the initiatives support the goals, but not just the goals of the group that created them. We must include the goals of all groups. Otherwise, we risk the danger of conflict and non-alignment.

6.3 How Work by One Group Impacts Others

Because each initiative has activities associated with it, let's look at this relationship more closely and see how they impact the goals. To do this we need to start at the Activity level. Activities have two main components. The first are the outcomes associated with each activity. The second is the timing of the activity from its beginning to its completion.

Outcomes

Everything you do has one or more outcomes that affect you and those around you. These outcomes are the results of executing a given activity. If you have planned and executed your work well, these outcomes will either directly or indirectly enable your activities and associated initiatives to be accomplished. In turn the initiatives affect the goals. Thus, activity outcomes influence goal outcomes.

Chapter 6

Activity outcomes also enable other indirect events to take place. Initially you may be thinking only about the work you are doing. However, Yyour activity outcomes can enable other groups to accomplish their work as well. That's the positive side. On the negative side, some outcomes can disable, causing failure not only in their own associated initiatives and goals, but also in the initiatives and goals of others. These enabling and disabling activities can determine success of any process. Figure 6-3 shows how these outcomes impact you and others.

In this figure, the *y*-axis looks at the outcomes from your work and how they affect you. They either enable you to succeed or disable your efforts, causing them to fail. But you need to look beyond their impact on

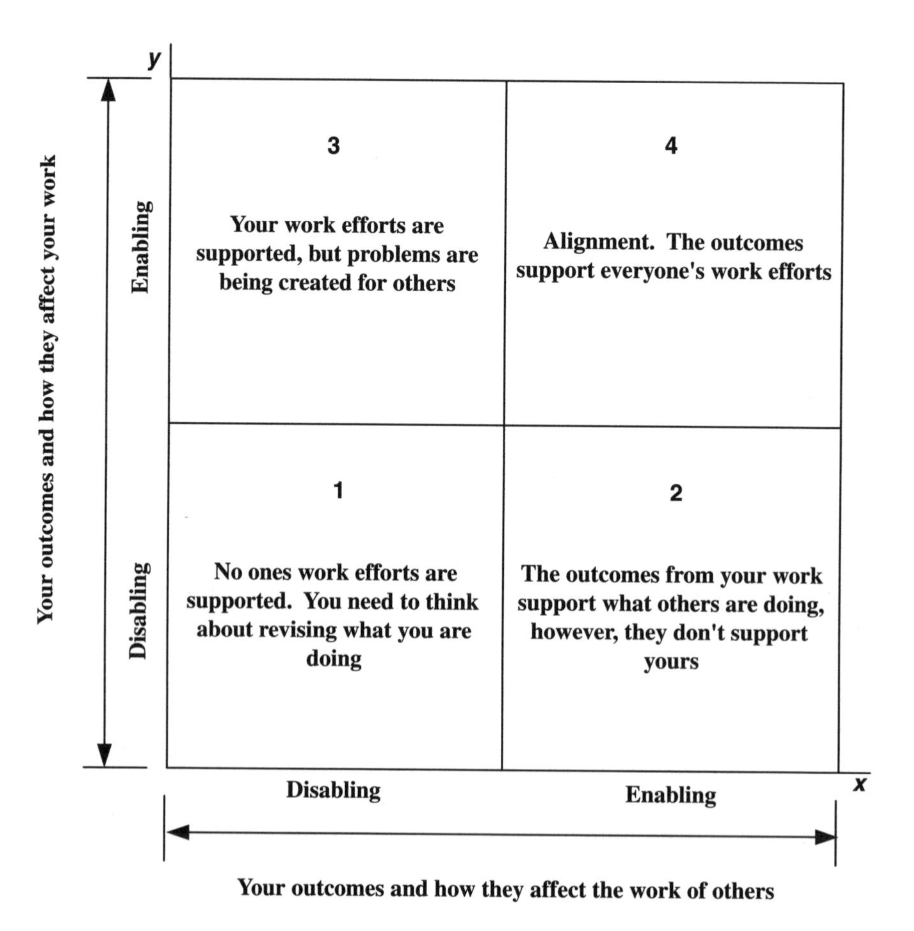

Figure 6-3 Your Outcomes and Their Effects on Others

you. How do they impact the ability of other groups to accomplish their work efforts? This impact is measured by the x-axis.

Quadrant 1—Outcomes are disabling to you and others. If outcomes are placing you in this quadrant, stop what you are doing and rethink the activity. It is bad enough that you are hurting others, but your own efforts are being negatively impacted as well. This quadrant is not where you want to be.

Quadrant 2—Outcomes are disabling to you but enable others. In this quadrant your outcomes support the work of others, but they don't support yours. You need to evaluate what you are doing and find out why this is happening.

Quadrant 3—Outcomes are enabling for you but disabling to others. Here your outcomes support your work efforts, but disable those of others. This problem can be overcome through some in-depth discussion with the other groups so that you can modify the activity and have outcomes that support you both.

Quadrant 4—Outcomes are enabling to both you and others. This is where you want to be. Your activities are set up in such a way that the outcomes support not only your efforts, but also the efforts of the other departments or groups.

TIMING

Timing is the second component to consider. Suppose one of your goals is to improve how work is executed in the plant. Flowing from this goal is an initiative to assemble a group to conduct a detailed work process redesign. Continuing along the Goal Achievement Model is an activity of gathering the pertinent data that will support the need for a redesign effort. If the team is going to meet next Monday, this activity has a time constraint. It is unacceptable to deliver the data in three weeks; you will have missed the deadline. Timing plays a very important role in the outcomes of activities.

As with outcomes, timing can positively or negatively impact not only your initiatives and outcomes, but also those of other departments and groups in your company. When timing is an enabling factor, there is no negative impact. It is when the timing of what you are doing is out of alignment with your activities or those of others that you need to address how you can correct and eliminate the problem. Timing is often more difficult to address. Most efforts have timetables that include steps that are either sequential or parallel as well as steps that are either dependent or independent of others. These factors add a high degree of complexity. Nevertheless, you must address these timing and alignment issues. A suggested starting point to solve this problem is to obtain a book on planning, specifically one that addresses the topic of bar charts (sometimes called Gantt Charts, named after the developer). Microsoft Project Planner can also be a useful piece of software for many, helping them plan their work sequences.

6.4 The Roadmap

Now that we have talked about outcomes and timing, we can focus specifically on the Roadmap. What the Roadmap tries to create for you is a way of assessing the outcomes from the activities. It is designed to help you identify the impact, either positive, negative, or neutralgative, that these outcomes will have on you or on others. You will also be able to identify whether or not any negative impacts are based on results or timing. related. The map also helps you to identify corrective actions needed to remove the negative impacts

Figure 6-4 zooms in on one portion of the Goal Achievement Model. It begins with one initiative and its related activities. It then follows the Outcomes of one of the Activities and, in turn, the Impacts and, if needed, Corrective Actions.

As you can see, the Goal Achievement Model provides a valuable overview, but only begins to scratch the surface of the effort shown in Figure 6-4. In order to adequately capture all of the relevant information, you actually need a worksheet for each expected outcome. Figure 6-5 shows a sample format for this worksheet. You may want to adapt it to fit your own needs content should remain the same.

6.5 How to Use the Worksheet

Using this process properly for each initiative and associated activities requires using many worksheets. For example, four initiatives each with four associated activities will require that you prepare at least sixteen worksheets, more if you use one worksheet for each outcome. Although this process may be time consuming, I truly believe this exercise is necessary. First, it requires you to think about each outcome and its impact on you

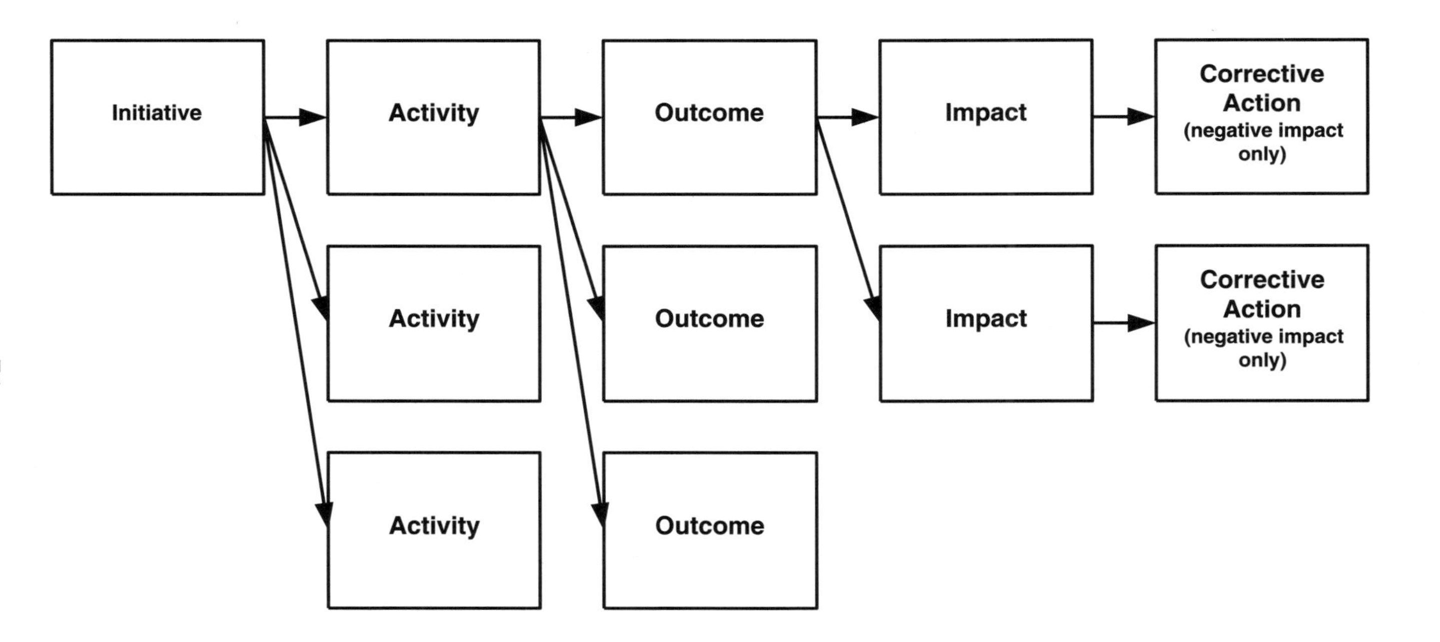

Figure 6-4 Goal Achievement Model Showing Outcomes and Impacts

Initiative:				
Activity:				
Expected Outcome:				
Impact:	On You	❑ Positive ❑ Negative ❑ No Impact	On Others	❑ Positive ❑ Negative ❑ No Impact
Based on:	Results Timing	❑ Yes ❑ No ❑ Yes ❑ No	Results Timing	❑ Yes ❑ No ❑ Yes ❑ No
Corrective Action				
Timing of Corrective Action:				
How Do You Know it Was Successful?				

Figure 6-5 Outcome / Impact / Corrective Action Worksheet

(This form can be found is on the disk at the back of the book)

and on others. Second, it forces you to follow a logical thought process. Only Quadrant 4 outcomes that positively impact both you and others can make this effort shorter. These outcomes do not require corrective actions

The first step is to select an initiative. Write it in the Initiative space on the worksheet. Next, select one of the associated activities to write in the Activity space. Each initiative activity combination will have multiple outcomes. Each outcome may have several possible impacts. Try to identify all possible outcomes and list them in the Expected Outcome section of the worksheet. Especially when you are first learning to use the roadmap, you should use a separate worksheet for each expected outcome. Some outcomes have positive impacts; others are negative. Check off on the worksheet whether the outcome has positive, negative or no impact on you and on others. (This information tells you which quadrant from Figure 6-3 describes the outcome)

Once this step is completed, you are through with those that are positive; Quadrant 4 outcomes are the results that you want. The negative outcomes are a different matter. For these you need to describe the corrective action you are going to take in order to change the outcome to a positive one. The worksheet provides a block for this description.

The last two steps involve identifying the timing for your corrective action and the way in which you will know that the effort was successful. Both of these sections on the form provide you with a method to track progress. It is useless to find problems, but not go through the process of correcting them.

Right about now you may be thinking that this process is a great deal of time consuming work. That may be true. But think for a minute how much more work is associated with fixing a change initiative after it has gone wrong. Wouldn't you rather do it right the first time and avoid all of the problems later?

6.6 A Detailed Example

Suppose that you work in a plant with a great deal of equipment and a large workforce to support it. The various computer systems that you have in place to support the work are old and developed by the site. They are not integrated with one another so that there are many sources for the same data. Currently you are redesigning the maintenance work process. In fact, this design appears as one of your goals on your Goal Achievement Model. From this goal, given the weakness of your current computerized maintenance management system (CMMS), you naturally develop an initiative that states that you will replace the current CMMS. In turn, this initiative leads to three activities: Developing a detailed CMMS work scope, selecting a qualified vendor, and installing the system in the plant.

Let's now focus on the first activity, developing the CMMS work scope. You create a team to handle the task. Using the Roadmap worksheet, you arrive at five outcomes, listed in Figure 6-6.

Activity: Develop a detailed Computerized Maintenance Management System work scope for the site		
Outcome	**Impact (+ or -) Who is Impacted**	**Results or Time Based**
The scope is developed per plant requirements. (Quadrant 4)	(+) Positive for all groups involved	Not applicable
The scope does not include reliability group and its software needs. (Quadrant 3)	(-) Negative impact on the reliability group	Results based
The scope provides less functionality for materials management than was present in the old system. (Quadrant 3)	(-) Negative impact on materials management	Results based
The timing for vendor selection does not finish until after the budget cycle. Funds can not be made available for next year. (Quadrant 1)	(-) Negative impact on your project and all other departments	Timing based
Costs are captured by the finance system but they do not show up in our CMMS. (Quadrant 2)	(-) Negative impact on your project (+) Positive impact of finance	Results based

Figure 6-6 Outcome – Impact – Results – The Computerized Maintenance System Example

Looking more closely at each outcome and its impact,

- You develop the required scope. This impact is positive; therefore, nothing further is required.
- The reliability group has needs that you didn't include in your scope. (In fact, you actually forgot to invite them to the meeting.) Here you have a negative impact that is based on the outcome of this activity.
- The scope provides the materials management group with less functionality than it has now. The group is not pleased. Here is another negative impact.
- The timing has you selecting a vendor after the budget cycle has been completed. If this timing is not changed, you will have no chance of obtaining funds for the work next year. This impact is timing based and is again negative
- The scope of work requires that the finance system currently in place be used for capturing the cost. As a result, the dollars do not show up in the CMMS. This result is also disabling. In this case, however, the result is disabling to you, while enabling the finance group.

Outcome	Corrective Action	Timing	Success Looks Like...
The scope does not include the reliability group and its software needs.	The scope needs be reworked to include the reliability.	As soon as possible.	The group's requirements have been included.
Materials management has less functionality with the new system than was present in the old system.	Develop ways to work around these problems or have the software modified.	During development of system at site.	No perceived loss of functionality via software additions or workarounds.
The timing for vendor selection is outside of the budget window.	Accelerate selection process so that funding can be included in budget.	As soon as possible.	System funded in next year's budget.
Costs are captured by the financial system, but not show up in CMMS.	Develop the proper interface.	Include with scope so that it is addressed in the bid.	Dollars show up in both systems.

Figure 6-7 Outcome – Corrective Action Outcome Corrective

In this simple example, we encounter five outcomes. Two fall into Quadrant 3, and one each into Quadrants 1, 2, and 4. The quadrants are identified in Figure 6-6. Figure 6-7 continues to summarize the worksheets, focusing on some of the associated corrective actions and timing for the negative impacts. As you can see in the figure, corrective actions were developed with associated completion times. The team has also identified what they feel success will look like when the corrective actions are completed.

The point I want to reiterate is this: It is far simpler to discover problems and develop corrective actions at the initial stage of a change effort. If you wait you may find out that, rather than adding value, you have created problems bigger than those you were trying to solve in the first place. Take the time up front; it is well worth your effort.

6.7 Pitfalls: How to Identify and Overcome Them

Now that we have talked about the Roadmap, let's look at obstacles that can hamper your ability to use it. Many pitfalls can impact your work plan and use of the model. They include:

- **"I already know the answer, so here is how we are going to do it."**
  This problem occurs when people in influential positions believe they already know the answer and are using you and your team just to validate it. The problem becomes obvious when there is a lack of openness for new ideas and new approaches. The team's suggestions fall on deaf ears. You can see the group becoming frustrated as it realizes that the client (management) knew the answer before the team was commissioned.

- **"You handle it and let me know the results."**
  This problem is the exact opposite of the previous one. In this case, manage ment only wants the final results. This approach presents a high likelihood that your presentation will not be what management wanted. In turn, a high level of frustration develops and people start dropping out of the process. To make matters worse, this problem usually gets communicated to the workforce through the "grapevine," leading many others to buy out as well. In the end, the approach sets up the entire facility for failure and creates a lot of anger among those who invested their time and effort.

- **The cube approach.**
  Think of a series of cubicles in which people get assignments, complete their portion, and then throw the work over the wall for someone else to handle. This is the cube approach. You know it exists when you ask others the status of a work effort and they reply, "I gave that to ______ and I don't know where it stands" or "I did my part and it's no longer my concern." A site working in this way has very little team structure. Resolving negative outcomes is very difficult.

- **The silo approach.**
  The silo approach is like the cube except, an entire department not just a single individual is included. Silos are functional in nature. Again, we are faced with problems when we try to resolve negative impacts that exist between departments.

- **The system will handle it.**
  Because we live in a world of computers and computer systems, most changes are in some way supported by systems applications. Many believe that if you simply install the system, the work processes and other concerns will take care of themselves. I guarantee you this is not the case. As you will see in Chapter 12, systems are enablers only. If you rely on systems to be the cure-all, you will not be happy with the results.

- **"I want it yesterday."**
  Everyone reading this book has heard this at least once, and probably more often. Many times, this response is indeed a necessary part of the business

process. However this type of response has no place in the world of change. As you progress through a change process, you and your team need time to do all required steps. Remember: In the world of change, slow is fast.

- **Just build the plan and follow it.**
  This model totally ignores spiral learning (Chapter 3). You cannot create a linear plan for a change process, then simply follow the plan. It won't work. Applying a linear approach will probably cause the work effort to miss many benefits that could be achieved if you were allowed to work through the spirals.

These pitfalls can get in your way as you use the process described in the last several chapters. Review them for a minute. What all of the pitfalls have in common is that they are beliefs that management thinks will get the job done. They are also in opposition to everything we have talked about.

What can you do to get senior management to help you and your team achieve success? One possible solution builds on how you manage your management. Yes that's right: how you manage your management. By managing expectations of those who are in positions of power and who consequently can support your change efforts, you have your best opportunity for success. Many projects have a segment during which participants list expectations of those who are being affected by the change. Few projects, however, have the additional component of recognizing that the support of senior management is a critical success factor and have established expectations for them as well.

In your own projects, start by creating a list of expectations for those intimately involved in the process. Then create a set of expectations for your senior management team as well. In this way, they will know what they need to do in order to make your effort a success. Furthermore, create this list early in the work effort. With their agreement to the expectation list, you can avoid at least some of the pitfalls just described.

6.8 Moving Onward

Chapters 4, 5 and 6 are really a package. Articulating your vision leads you directly to the Goal Achievement Model. This model, in turn, prepares you to use the Roadmap. Each of these is a tool that enables you to find a new way of developing an idea and then turning it into reality. Many people who have great conceptual ideas fall short in the ability to convert their ideas into something substantial, something that a business can use to improve. The models described in the last three chapters will help you through this process. As I have already noted, this process is not easy. It demands hard work for you as the change leader and for those members of your team who have been asked to undertake this task.

Chapter 7
A Case for Teams

7.1 An Introduction to Teams

This chapter is about teams. The focus will be how to make teams work, either where they have never existed before or where they currently exist, but not in the way that creates or adds value to the organization.

First we must establish exactly what a team is and why teams should be used in everyday business. In The Wisdom of Teams (New York: Harper, 1994) Jon R. Katzenbach and Douglas K. Smith define a team as follows:

> "A team is a *small number of people* with *complementary skills* who are *committed to a common purpose, performance goals, and approach* for which they hold themselves *mutually accountable.*"

I have underlined five key aspects of this definition, all equally important, and for this reason I want to spend some time discussing each one separately.

"A small number of people"

Some people think that the size of a team does not matter. They think that if you want people with a common purpose to work together, then you simply need to get all interested parties together for maximum benefit. Yet size is important. If a team is too small—less than four—it probably will not be of sufficient size to represent all of those who would be affected by its work. Furthermore, such a small group most likely will not have enough membership to accomplish the task. On the other hand, if a group is too large—greater than twelve—it could be too large to get anything accomplished. My own opinion is that the optimal size for a team ranges from four to twelve. Size is important to consider when you are creating a team because you want positive results, not frustration and failure.

Sometimes, getting the right number of people on the team is very easy. However, you may run into the problem of not having enough or having too many members. What can you do? If you don't have enough members, consider getting two or three representatives from each affected area in order to increase the team's size. If this solution isn't workable, then consider geting people from other groups that may not be as directly affected, but that still have a stake in the issue. Use them to increase the numbers. When you are short the number of members you need, then check that you have included all groups that are affected. Often the issue of size will resolve itself when you recognize that you have inadvertently left out a group that should have been included.

Shortage of team members is usually not the problem. The problem usually is the opposite: having too many people for the team. Often when you are working on an initiative that will affect your company, many groups need and want to be included. The result can be that you have a team of more than twelve. You can apply two possible solutions to this case. The first is to divide the problem into parts, then create more than one team. The benefit is that you have groups of manageable size. The risk is the possibility that the smaller groups arrive at conflicting solutions. Bringing the groups back together on a periodic basis to validate what the others are doing can mitigate this problem. You can also have group representatives meet and agree on a common direction, thereby minimizing this risk.

The second solution is to start with one large team, then break it into smaller subgroups to work on parts of the problem. Periodically the small groups are brought back into the large team format. Consistency of approach can then be maintained.

In the first of these approaches, you start with many small groups and bring them together to work out issues. In the second, you start with one large team, then break into smaller groups to do the work. The first approach can be used when consistency between the groups is not a major concern, the second approach when consistency is important.

"Complementary skills"

Another important aspect of a team is that the members have complementary skills. The collective skills of the members must provide what you need to accomplish the team's purpose. The old adage that the whole is greater than the sum of the parts is most certainly true in successful teams. Every sport played by a team demonstrates the power of complementary skills; individual skills that when taken together create a team that can accomplish the desired outcome.

"Committed to a common purpose"

If a group of people is trying to accomplish something that they collectively want to achieve, then they are committed to a common purpose. The area where a problem most often occurs is in the definition of "commitment." How do you define or know the level of commitment of individual team members? How do you determine the level of commitment that is necessary? These questions are not easily answered because commitment comes in varying degrees. It is determined by the team and, in most cases, defined by the team's specific assignment at any given time. A good team is self-regulating. If it is a mature team, its members will identify the level of commitment that is required.

Suppose a team is working on a critical business problem. The team must meet four hours per day for a week. People who are late for the meetings or don't provide the necessary level of input are likely to have their behavior addressed by the other team members. The group needs must be fulfilled. At times, a request for additional commitment may be made in a joking way. Sometimes an outright demand for more commitment may be necessary. In the end, the team will try to regulate from within the needed commitment from its members.

"Performance goals and approach"

A team can not be a team unless it has something to do. Business teams usually have goals or initiatives and activities on which the team can focus its attention. Although the team may start at the goal level, it quickly moves through the stages of the Goal Achievement Model and actually works at the initiative or activity level. It is here where teams can create value for your company.

The specific approach that a team takes will vary depending on the assignment. In all cases, they must follow a common approach within the team so that their time is optimized. Attending a meeting with no agenda and no process, or one where the meeting doesn't flow, frustrates team members and wastes their time. A good team does not allow this to happen; it is focused on results.

"Mutually accountable"

Mutual accountability is another factor important to a team's success. Certainly teams and their members are accountable a higher level within the organization. Equally important is the fact that the members are accountable to each other. Each member brings a separate complementary skill to the team that requires the members to depend on each other.

This aspect of teams also points to a potential problem. Teams are often built across functional organizations; the members are from different departments and work groups, and even from different levels of hierarchy. As the individual members go through the stages of becoming a team, the members bond; they become mutually accountable to one another. When the team's assignment is a project with a finite end point, then the problem of accountability can be minimized. By the time the members get to conflicts between their own regular responsibilities and the team's, the project is usually coming to an end.

What about the long–term efforts? Teams that operate continuously effectively create a matrix organization within the company. Thus you would be part of a regular work group and have an immediate superior. At the same time, as a member of a team, you may be supervised by someone outside of your immediate organization. In this sense you have two bosses. This is not a problem when the goals and initiatives of your immediate supervisor and the team are in alignment. But what if they are not in alignment? This conflict can create a problem for your personal performance, your work group, and the team.

When you are pulled in two different directions, which way do you go? Unfortunately for teams in this situation, the members usually align themselves with their immediate supervisor, the one who pays their salary. In turn, problems develop for the team, inhibiting its ability to do its work. This conflict speaks to the need for alignment. Then what you do for the team will also complement what you do for your primary job and work environment. Without alignment something will suffer—and it usually is the team.

7.2 Types of Teams

Teams can be divided between those that work on projects that have a distinct beginning and ending and those that are established to provide ongoing support for a change initiative. The second type generally has no end.

The life of a team working on a finite project can be as short as one meeting or it can last several years. In either case, when the work is completed, the team can disband. These project teams are an important part of the change effort. They effectively implement change. Working from the vision and goals, they develop the design of the change effort. It seems to me that these project teams are needed to initiate and implement a major change in an organization.

The second type of team often results from the change effort you put in place. These teams become part of the structure; they are integral to the work process. These process teams monitor the organization to ensure that any change is maintained. When the project teams have completed their rollout of the initiative, the process teams stay, becoming an integral part of the new organization.

If you are serious about change in your company, then you will need a project team to design and implement the change. In most cases, you would then need process teams to maintain the change initiative.

7.3 Why Teams are Needed for Change to be Successful

At the minimum, a project team is needed to institute a change initiative. I also believe that for any step changes, process teams will be needed to sustain and even grow the change initiative over time. Process teams may not be needed for small incremental changes but they are important for major change efforts. If we want success, we can not continue to use the old organizational structures. What, then, is the difference between the team concept and older types of management style? Furthermore, why should we change?

Chapter 7

The real manufacturing boom in the United States started in the years after World War II. At that time, industry was growing by leaps and bounds. Labor was not expensive. Compared to today, a great many manual tasks needed to be coordinated across large organizations. Given a workforce that, by today's standards had relatively little education, there was a strong need for tight control and communications across the various work functions. The organizational model that best fit this scenario was the "command and control" model of the military. The model had worked extremely well during the war. Its basic premises for running a business with a lot of people performing individual, low-level, yet integrated tasks fit very well.

Over time, American business changed. With increased competition, companies began to automate the work. As a result, cuts were made in the workforce. Even during this time, the military model still worked satisfactorily. However, one major component changed all that. As a result of the G.I. Bill, the growth of public colleges, and the emergence of the community college system, the workforce was becoming more educated. A higher level of training was also needed to run the equipment in the increasingly automated plants.

As productivity increased, workers needed to perform varied tasks. Because of their education, they were ready for the challenge. However, the existing command and control model had produced workers who expected to do what they were told, without thinking a lot about their assignment. This model was in conflict with the increasingly automated plant that required more responsibility and thinking from the employees. In addition, employees wanted to be a valued part of the business. They began to resist the command and control model. Over the past three decades, we have witnessed the gradual but steady demise of command and control approach in the United States. It didn't die without a fight, and it still exists in many places today, however not to the extent it did before.

I've never forgotten the time I watched a foreman, years ago, tell one of his mechanics, "I didn't ask for your opinion, just do what I told you." The mechanic may have had a good idea, but he was cut short. Any input he may have had was dismissed as irrelevant. He did in fact have a good idea to offer. He was highly skilled, not just in his craft. Outside of work, he was the mayor of a nearby township.

Among the replacements to the old structure was one that emphasized teams and their many applications to business. Today's businesses require that many parts of the organization work together. Teams address this requirement rather well. Many of the work process problems in business, especially those based on cooperation and communication between departments, result from a lack of teamwork.

Many years ago, the plant where I work purchased electrical generating equipment. During the summer months, the power company would enact "brown outs" (voltage reductions). The generating equipment could handle the electrical loads so that we wouldn't have to reduce production. But how were we going to operate the equipment? It didn't run continuously—only during the day shift during the infrequent "brown out" periods. Start up-times were random. Because we were a continuous process plant, we operated with rotating shifts. However, because of the random need for start up, we didn't know which shift was going to be on duty when the power company asked us to turn on the generator. In addition, our operating crews were not of sufficient size to enable some of them to leave their primary job for power generator duty. With these problems facing us, we had a hard time figuring out how to run this equipment.

In a command and control environment, the plant would have mandated training for a very large group of operators. It would have expected them to operate when ordered to do so. Yet no one would have a great deal of operating experience with the equipment. This inexperience would have led to major problems.

Fortunately these problems were avoided by the use of teams. With the help of our union, we solicited maintenance mechanics to own, operate, and maintain the equipment—a true team approach. This was a different approach for plant employees who had worked primarily in a command and control environment. We asked people whose prior experience was "do what you are told" to learn how to operate the equipment, maintain it, and, direct their own schedules and minor maintenance activities. This approach represented a major change. But we wanted to use the whole person. As a result, we exceeded our goals beyond our greatest expectations. We had created a team that operated and maintained the equipment, a phenomenon that had not previously existed at the plant. Furthermore, the team members developed a cohesiveness that I had never seen before. In addition they saved the company a great deal of money. Prior to this experience, I had only heard and read about teams. Here was real evidence in the work place. I have never forgotten it.

Chapter 7

7.4 Team Readiness

Project teams which will be discussed further in Chapter 14, are formed to identify and implement change. Process teams are designed to become an integral part of how a company conducts its business.

Process teams are special. They need to be ready before the change process can be put into place. Consequently, readiness is a plant-wide issue, not just a concern of a team that has been selected for a specific task.

Chapters 11 and 12 look at eight elements that are central to any organizational change: leadership, work process, structure, learning, technology, communication, interrelationships, and rewards. These same elements also determine if a site is ready to undertake a team-based initiative. Therefore, let's introduce these elements and use them to determine if a site is ready for change.

Leadership. Do the leaders of the site support a structure where they do not have all of the power? In a team environment, management has to relinquish a certain amount of power to the teams. Otherwise, the leadership is not ready to support the team process.

Work Process. Does the site's work process function in a way that promotes collaboration? Part of almost any redesign effort is focused on changing the work process. The current process can tell you a lot about how receptive the employees will be to teams. If employees complete their part of the process, but do not feel any ownership for the whole effort, then team readiness may be suspect. On the other hand, if they pass the work along, but still feel ownership for the total product, teams may be far easier to achieve.

Structure. Is the organization fixated on working things through the chain of command? Such a fixation would be evidence of a lack of team readiness. Another indicator is the interest and ability of management and employees at all levels being willing to work together. If they are not, then team readiness does not exist. It is not always the formal structure that makes a site ready for teams. The internal structure—how things really get done—is equally important and needs consideration.

Group Learning. Group learning is not the same as training where individual or group skills are enhanced. Instead it is that quality of a site that permits employees to gain knowledge from the events that have taken place, then apply this knowledge to the future. If teams are to be introduced, the organization needs to be able to learn and develop from its experiences. A good indication of a learning organization is one that reviews an adverse situation and asks, "How can we prevent this from happening in the future?"

An organization that has not developed learning skills is often focused on finding someone who can be punished for the failure.

Technology. By supporting the work processes, communication, and data that teams need to be able to make sound, data-driven decisions, technology and information systems bring strength to team efforts. Using technology and having a clear understanding that it is not a driver, but an enabler of improvement, are good indicators of team readiness.

Communication. In order for a site to be ready for teams, a good communication system must be in place and those at the site need to use it. Examples include Voice mail, e-mail, plant publications, and related announcements. Having these systems is not sufficient. Having them without using them is the same as not having them at all. A second component of communication must also be in place as part of team readiness: the structures and processes that enable, and even encourage, communication across the levels of the organization, both up and down, and across departments.

Interrelationships. An organization has many vertical and horizontal structures. The strength of the personal and working relationships among these structures, as well as with the unions in organized plants, is an indicator of team readiness. For teams to function well over time, these relationships need to be mature and non-adversarial. Without the strength of these more informal relationships, some of the glue that binds teams together will be missing.

Rewards. A plant or company that is ready for teams should be evolving to team-based rewards. Rewarding people in this fashion runs contrary too much of our collective upbringing. Most of us are taught the value of individual achievement. Although schools are beginning to institute teamwork efforts, they are still in the minority. With team rewards, you are rewarded based on the results of the team's efforts, not your own individual effort. A performance appraisal system in which a manager gets input from subordinates and peers is a good indicator that a site is ready for a team-based reward structure.

The degree to which a site has incorporated these eight elements helps to measure whether or not it is ready to move towards teams as a way of conducting business. Not many sites have mastered all of them and are functioning in a way that makes the sites ideal for a team-based work environment. However those that will be successful in instituting teams will have mastered many of the readiness components. They will also be working on, or at least aware of, any gaps in those that remain.

Chapter 7

7.5 Where You Are and How You Get to Teams

Assume that you and those at your work site have decided that teams are an integral part of a work process redesign. You must strongly believe in teams if you want them to work. Otherwise, if you are just going through the motions, the people involved will know it and the team initiative will not work.

However, if you do believe strongly in teams, then you are at one of three places. One is that you don't have teams, but want to have them. Another is that you do have some form of teams, but they are not working well. The third is that you have teams that you think work well and you are nodding your head in agreement with my last statement. If you do have teams that work well, then you may consider yourself very lucky. However, it wasn't really luck at all, was it? It was actually a lot of hard work over a long period of time in order to become successful. The movement to effective and efficient teams can not be mandated, nor is it an overnight event.

We Want Teams But They Don't Yet Exist

For teams to exist, you need to be ready. Teams can not exist simply as an addition to what you already do; they need to be fully integrated to the point that the process requires their existence. If the process is set up to require teams, then the actual move to teams will be a compelling driver for the organization. In this case, it will actually be more difficult to remain with the old process than change to the new.

Suppose you have redesigned the work process. Part of the design indicates that teams would be a good addition. However, you never provided a strong team structure or a real reason for teams to exist. For those who see the value of teams, the teams may take hold. But for those who prefer the old way, teams will fail. Even though some people may believe that teams are important, teams have not been made a compelling and integral part of the process.

To make teams work you can not just tell everyone you want teams. You must build your change process so that people at the work site can not exist without them. You must also create a set of expectations based on the value of healthy functioning teams. If you successfully create both the change process and the expectations, then teams will form. Your job then becomes to feed them and make them healthy.

Chapter 7

We Have Teams But They Are Not Working Well

Sometimes teams exist but they are not working well. Your goal is to turn them around so that they become a sound part of the foundation for your change effort. In this case, review the eight elements needed for team readiness. Use them as a troubleshooting guide because something has gone wrong somewhere in the development or use of the teams in your effort. Once you find out where the problem is, you can correct it. Only then can teams change from the ineffective ones that you have now to ones that will play an integral role in the work effort.

Suppose you work in a plant of the XYZ Company. The teams that were built into the change process are not working. The new structure provides people with different skills to each of several operational teams, but the organization itself is still basically built along functional lines.

You look at your process and how teams fit into it from the standpoint of the definition of teams In this example, suppose these are the results of your review:

- Small number of people. In order to have the right skills, the teams do have a small number of people.
- Complementary skills. The skills necessary for the operation of the team have been provided. Due to resource issues in some departments, however some team members are on multiple teams, reducing their effectiveness.
- Committed to a common purpose. The team's common purpose—to operate in a way that ensures optimum utilization and maximum profit—is loosely defined. Thus, the purpose means different things to the different members.
- Performance goals. Because the functional departments still play the primary role in the plant, people's performance goals are focused on departmental and individual goals instead of being focused on success of the teams.
- Approach. The site said teams were important. However, much of the work process is still focused on departmental responsibility. The teams do not have assignments specifically for them. Although parts of the process should be handled directly by teams, management has not made this assignment.
- Mutually accountable. The team members have complementary skills. Yet much of their work is departmental, not team focused. Therefore, the teams often have trouble accomplishing tasks because their members are working either with other teams or for departmental supervisors who have directed their priorities away from the team.

Based on this analysis, it is fairly obvious where the problems exist.

- A small number of people comprise the team, but they have other responsibilities that dilute the skills they bring to the team.
- The team has no common purpose that binds its members into a cohesive unit.
- Performance goals and the approach to the work are departmental instead of team based.
- The departmental focus detracts from the accountability of members to the teams.

Taking a closer look at the eight elements of team readiness can help to develop the solution to these problems. From this review, you can determine what adjustments are needed to correct the deficiencies within in each. For our example, I have focused on five of the eight elements because here is where the issues exist.

- Leadership. The site needs to provide a common purpose or reason for wanting teams. Teams were created to improve the business, yet they were never given tasks that required their existence. Without specific assignments, the members focused their attention on other tasks at the expense of the team.
- Work Process. Although the work process has a team focus, it is not as important as the functional role. For success, the team focus must have equal or greater importance.
- Structure. The structure of the teams was not designed for long-term success. Members have multiple team responsibilities as well as responsibility to their original functional areas. As matrix organizations show, teams can exist in this type of environment, but their relationships need to be carefully crafted.
- Interrelationships. There is no direct evidence that problems exist in the relation ships among team members. However the departmental focus is likely to cause problems, leading to the breakdown of the team structure. It is very frustrating for a team to be working on an assignment, but have one or more members unable to participate because they are working on departmental tasks for their supervisors. Their team efforts are pre-empted, causing tension among the team members.
- Rewards. The performance goals and, in turn the reward structure are not team based. Management needs to put into place a reward structure that provides greater emphasis on team performance.

Looking at this example, you can see how to make teams more effective. Start with the team process already in place, analyze it against the definition of teams, then use the eight readiness elements to develop strategies for correcting the problem. Taking an organization that has an ineffective team structure and then making the changes needed to make teams effective is very difficult. You need time and commitment from all levels of the organization.

Chapter 7

7.6 The Concept of Critical Mass

If we believe that teams are the way to go in order to implement lasting and beneficial change, then another question arises: how many people need to buy into the concept before it takes hold? The answer to this question does not apply only to teams. It actually applies to any change initiative. Initially someone answering this question might say you need 100 percent participation to achieve success. Fortunately the change process does not need that complete amount. What you do need is an amount that is often referred to as the critical mass: The critical mass is the number of people who, once invested in the effort, will give enough credibility that everyone else will follow. Is the critical mass the majority of the participants? Sometimes it is, but often it can be less than the majority.

People will model their leaders if they respect what they stand for, what they do, and how they act. Sometimes leaders are the formal ones who head up the teams and departments. Sometimes however, they are informal leaders who exist in every organization. Once these leaders have been brought into the change effort and are modeling the correct behavior, the rest will follow. These leaders and those that follow their lead provide the critical mass.

When you build teams, you should identify the members who comprise the critical mass. Make certain you get them to buy into the change you are promoting. Once you establish the critical mass, the others will follow.

7.7 Facilitators. A Value Added Resource

Facilitators are an important resource who can help the team process be a success. They often have no technical knowledge about the content of the team's assignment. Their role is something else altogether. Their real purpose is to monitor the team's work process. They observe how the team functions, then provide objective insight that helps the team learn to work better together. In order to keep their focus on the team's work processes, it is generally better that facilitators have no technical insight to offer about the assignment. If they did, they might focus on the work itself rather than the process.

Many companies make the mistake of putting a person into the group who has both facilitator skills and content knowledge. People who tend to have good facilitation skills are often senior managers who have the content knowledge as well. But if this person doesn't clearly understand his or her role and does not allow the team to function independently then problems will ensue.

An alternative approach is to select someone from within the organization that has facilitation skills, but no applicable content expertise. Another solution is to hire someone from outside the company. You could also select someone from a consulting firm that your company uses. However, the last two options present risks unless you specifically know the skill level of the person you hire.

Because the other alternatives are not easy to achieve, the team must usually go with someone from their organization that has both facilitation skills and content knowledge. In order to avoid the problems just described, you should develop a facilitator contract or list of expectations. This contract or list briefly describes what you expect the facilitator to do. It should also include what you expect the facilitator not to do. My own list of facilitator do's and don'ts follows. You may want to add or delete your own items. Nevertheless, you need the team to review and agree to this document before you bring the facilitator into the picture. Once the facilitator starts, you need to frequently review the document and make sure that they understand their role in the effort. Additionally, with this contract, the team can help the facilitator stay on track should they wander off course, just as the facilitator is doing for the team.

Facilitator Dos

- Focus on team process, not content.
- Monitor team logistics, including agendas and meeting notes.
- Control the use of time – address lateness or non-attendance concerns.
- Objectively address team process issues that are not working, either with the team or with individuals.
- Ask questions that help the team to address issues in a comprehensive manner.
- Don't allow the team to be swayed by a dominant member.
- Make certain everyone is involved. See that everyone participates.

Facilitator Don'ts

- Don't provide content to the team unless asked or you have something of significance to offer. (In either case, ask for permission to change roles. Then provide your input as a team member and go back to facilitation.)
- Don't try to run the team. That is not the facilitator's role.
- Don't confront issues involving individual action in front of the team. (Provide corrective action in private and praise in public.)

Chapter 7

This list is brief. If you're interested in being a team facilitator, look for books that focus strictly on how to facilitate groups. While your team may not need facilitation on a continuous basis, you should strongly consider it when you start. A facilitator adds tremendous value, they are well worth the time, effort, and training cost.

Chapter 8
Working with Consultants

8.1 What Is a Change Management Consultant?

If you are involved in any form of organizational change, then you have probably worked with consultants on more than one occasion. This chapter looks at consultants and the services they provide. Armed with this information, you will be able to make intelligent choices as to whether or not you will use consultants to support your change effort and, if so, in what capacity.

I have worked with consultants for many years on various change-related initiatives. If you are like me, and are trying to make a step change or improvement, you will probably need their services. The trick is to pick the right consultant offering the most beneficial services. The right consultants will add value. The downside of selecting the wrong consultant is that they will set the effort back. Even worse, they could even destroy what you have worked to achieve.

Before we can look at how to select and best use change management consultants, we need to clarify what they are and what they do. Change management consultants are individuals, groups, and sometimes even entire companies with an expert level of knowledge that they can use to support your change initiative. Their support includes;

- Assessing your current work process and identifying performance gaps
- Providing expert advice about what other companies, both in and outside your industry are doing to improve
- Developing a vision and a model of your work processes in the future
- Preparing a detailed work plan or roadmap for change
- Redesigning the work process, either independently or in conjunction with personnel at the site
- Helping move the process forward in a planned and organized manner
- Leading and managing the project
- Providing resources to support the work effort and supplement those of the site
- Evaluating and implementing software that supports the process
- Developing and delivering training

Not every consultant can offer every one of these types of support. Nor will every consultant be an expert in your area of change. Therefore, you need to carefully evaluate consultants before you select one You need to identify which services you need, then find one or more consultants who can provide these services. As you evaluate the consultants, consider the type of company they have, their areas of expertise, and their philosophy for conducting the work.

Type of Company. At the small end, consultant firms can be individuals who are self employed and have expertise in a specific field, often in a specific industry. At the other end are large consulting firms with thousands of employees, addressing many areas of the work process. Some firms operate in a niche market, focusing their efforts in a key area. These firms are usually the small to medium sized firms, groups, or even individuals. Because of their lack of multiple areas of expertise and lack of depth of resources, these smaller companies are often somewhat limited in the size of the project they can undertake. However, they may be more cost-effective, depending on the size of your project, and offer more personalized service.

Areas of Expertise. A consultant firm can not be everything to everyone. It usually has an area of expertise. That area may be a specific type of industry such as oil and gas, paper, or power. Their area of expertise may also be a specific function within the industry, such as maintenance, materials management, or finance. Single-industry specialists are fine if you can find one that has expertise specifically in your industry.

Function-specific firms are also desirable if you only want to improve a specific function. A word of caution: Most work efforts that require consultants may start by addressing a specific function; however, if you really are redesigning a work process, these efforts can quickly become multifunctional. By limiting your effort to a specific functional area, you may quickly run into problems if the consultant is not capable of working beyond that area.

8.2 Work Philosophy

The work philosophy of change management consultants is also important. If their philosophy is not compatible with yours, then you may have serious problems.

The answers to three questions help determine your compatibility. First, what does the consultant want from the work effort? What will provide them satisfaction? Second, in the same vein, what do you want from the work effort? Third, what do you and the consultant want from your working relationship?

What Does the Consultant Want?

You may think that money is the only driving force for the majority of consultants. However, there are other factors. If a consultant were solely interested in money, then some of their decisions and recommendations would never be made. Self-serving actions by consultants would make them more money but the clients would not receive the optimum benefit. If this were the case, many clients would have a negative view of consultants, damaging their reputation and making it more difficult for them to get more business.

I think that the good consultants are motivated to improve the state of their clients' businesses and work processes. Their experience then expands their knowledge in the functional areas that they address. With these successes, come referrals, an expanding business base, and ultimately revenue.

What Do You Want?

You have most likely sought a consultant for the following three reasons:

1. You want your change initiative to be successful.
2. You want success the first time. Management is often unforgiving when it comes to false starts. In addition, if you are not successful with change initiatives the first time, the organization may not give you a second chance.
3. You want to optimize the change effort in the areas of schedule, resources and cost.

The drivers for these include:
1. You don't have the experience leading a change effort.
2. You don't have a broad base of functional knowledge other than your own experience or that of your co-workers.
3. You have other responsibilities. Therefore you do not have the time, nor are you provided the time, needed to manage the work process change over an extended period.

How do you reconcile your requirements in the first list with the drivers in the second? The answer is to hire consultants. They can bridge the gap between your business drivers and your requirements. By bridging that gap, they help you accomplish your goals.

Mutual Relationships

Knowing what the consultant hopes to achieve and having a clear understanding of what you want to achieve leads you to the task of finding a consultant whose work philosophy matches yours. However, I am not referring to the consultant's knowledge or expertise as it relates to subject matter. Instead I am describing how you and the consultant want to approach the work. You approaches must be in harmony for the effort to be successful. Specifically, you should look at two key elements: involvement and control. To what extent is the consultant involved in the project? To what degree does the consultant control the project? A bad mix of control and involvement by either party will be disastrous for the change effort. For example,

- Both the client and the consultant want the other party to control the project. As a result, there is minimal control. The project flounders and, in the end, fails.
- The client over-controls the consultant. As a result the process yields little or no value. All of the consultant's valuable expertise is lost.

- The client abdicates control and is not involved. The consultant, by default, takes over and runs the project. As a result the effort is successful only as long as the consultant is onsite. As soon as the consultant leaves the effort fails. Why would you expect otherwise? The owners are no longer present.

You can think of the work philosophy as a control–involvement matrix. In Figure 8-1, the x-axis is the level of control and involvement that the consultant has over the project. The y-axis is the level of involvement and control that you (your company) have over the project.

Let's examine each of the quadrants in Figure 8-1 to gain a better understanding of the involvement–control relationship between you and your consultant.

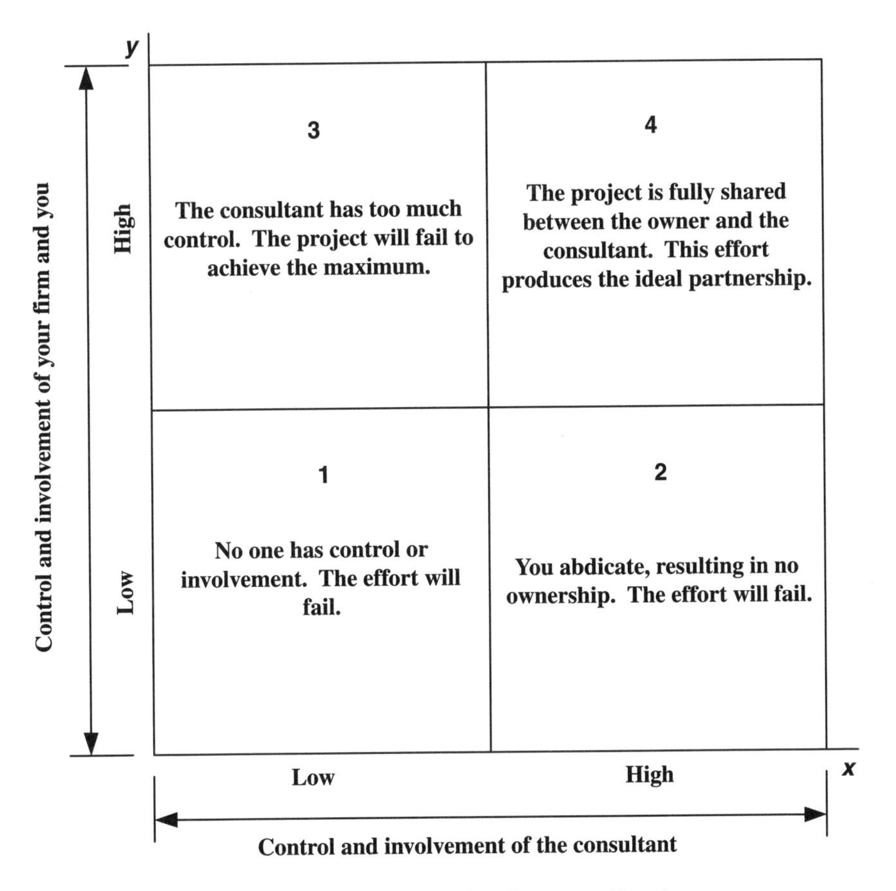

Figure 8-1 Control – Involvement Matrix

Quadrant 1—Involvement and Control of both you and the consultant is low. Good consultants do not actively seek low involvement and control. They are aware of the balance required between consultant and client for a successful effort. Consultants who end up with low involvement are often responding to clients who initially had good intentions when they hired the consultant but then cut costs. What happens is the client begins to see that a successful change effort with the proper level of consultant involvement is expensive. In such cases, the client works with the consultant to reduce the cost. Because consultants basically sell their knowledge and skills, cost reduction usually leads to reduced scope and reduced resources. Even though you hired the consultants for these specific attributes, as the client you have lowered their involvement and control. Additionally, you have dramatically lowered the value that they bring to the work effort.

At the same time, if the client doesn't recognize the significance of the change requirements on its resources and fails to appoint a fulltime team, then it has further lowered the likelihood of success. With both client and consultant having low involvement, the project is doomed. If this approach is taken at the outset, or if you see this approach being applied as the project evolves, stop the project until you can correct the approach before the disaster strikes. Unless you correct this problem, the change effort will be less than successful. Even worse, the desired change may never take place. Additionally, the organization will have little or no tolerance for change in the future.

Quadrant 2—The consultant has high involvement and control, whereas yours is low. This combination occurs when you and your company do not appreciate the significance of your role in work process change. As a result, you do not see the need for a fulltime team, even though one is required. Assuming you haven't followed the Quadrant 1 approach, your company is willing to spend money on consultants, using them to fill the voids in your own change team. The good consultants will see the flaw in a Quadrant 2 approach (lower client involvement). They will probably even tell you about it. Having that information and acting on it, however, are two different things. If you still choose to let the consultants fill this void, they will. As a result, you abdicate your responsibility for both the effort and the outcome.

The real problem does not appear until the work effort has ended and the consultant leaves. In the Quadrant 2 approach, you don't own the change process; the consultants do by default. When they leave your site, the change effort goes with them. Good consultants will try to prevent this problem by developing a sound transition plan that transfers ownership from them to you. However, you must be willing to accept ownership. The question becomes: If your firm didn't take ownership during the work, will they take it on at its conclusion?

Quadrant 3—The consultant has low involvement and control, whereas yours is high. When this combination occurs, your company has realized the need for a dedicated project team and the company team has accepted responsibility for the effort. This is good. The downside of this quadrant is that the company has allowed the consultants little involvement or control. This approach lowers the cost, but also lowers the resources and knowledge applied to the effort by the consultants. In most cases, this problem is caused by the client. However, the consultants may have overextended their resources, also leading to a resource and knowledge deficit. The result of this approach is that the change will have little, if any, effect on the existing process. With little consultant input, the company team will tend to replicate what is existing and, at the minimum, make only incremental changes when a major step change is required.

Quadrant 4—The consultant and you both have high levels of involvement and control. This combination is the best of the four approaches. Your company has recognized the need for a dedicated team. You have realized that the correct amount of consultant time and effort, while not being inexpensive, adds value to your business. This quadrant represents the partnership approach. When the consultants are finished with this type of initiative, you have site ownership. What was created will be perpetuated.

The optimum approach to a consultant – client relationship is Quadrant 4. You and the consultant collaborate to achieve success for your firm, while allowing the consultants to achieve their objectives. The balance is difficult to achieve. For a change process to be successful, however, the balance is necessary and you need to be aware that it is your goal from the outset. All of your relations with the consultants—from initial selection, through the project, to the time they leave the site—must be driven with a Quadrant 4 approach.

8.3 Consultant Projects

We have now identified the motivation for both the client and the consultant, and looked at how a collaborative effort has the best chance of success. We can now consider how the project and the organization's experience can affect the client-consultant relationship. Consultants undertake many types of projects, from simple reviews and minor process modifications through major change efforts. For any given project, there is a close correlation between the consultant's level of entry into the organization and the maturity level of the client site. The maturity level represents the experience of the organization working with consultants in the area of change management. Figure 8-2 illustrates the correlation.

In the figure, the *x*-axis measures the organization's experience working with consultants and change projects. Here a low level of maturity is represented by 1, a medium level by 2, and a high level by 3. The *y*-axis measures the level of the organization at which the consultants need to interact in order to be successful, with a representing high levels, b medium levels, and c bottom levels.

To determine the best entry level for the consultants, identify the organization's maturity level on the *x*-axis. Next draw a vertical line until it connects with the correlation line and then horizontally to the *y*-axis. The y-axis provides you with the correct entry level.

At the top of the *y*-axis, consultants first work with senior plant management before moving lower into the organization. This level of entry usually correlates to new or start-up projects. By starting major projects at the top, the entire organization can see that the project is supported. Through proper communication and development, those lower in the organization are then comfortable buying into the change. Major change projects rarely originate at the bottom of an organization, unless the organization has a high level of maturity for change. For new projects that include a lot of change, this is seldom the case.

As the organization's experience with change increases, we move to the right along the *x*-axis. You can see this by noting that a project for an organization with low maturity (point 1) requires entry at the top (point a). A project for an organization with high maturity (point 3) permits entry at a lower organizational level (point c). Obviously these are the two extremes. There are varying levels of organizational maturity in between that would permit the consultants to enter somewhere in the middle of the organization. This is shown by the points b and 2.

Before you bring consultants into your organization, determine their correct entry point. Allowing them to enter at the wrong level is a pathway to failure. Suppose your company decides that a major change initiative should be a grass-roots or bottom-up effort. In this case, the con-

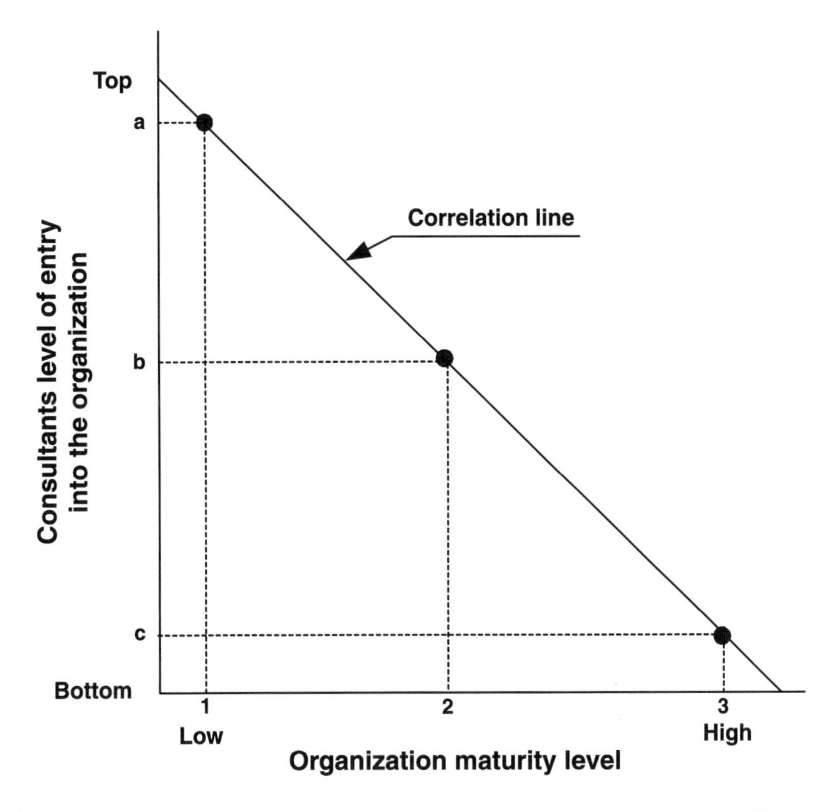

Figure 8-2 Organizational Maturity and the Level of Consultant Entry

sultant might enter at point c. However, your organization has not been exposed to change efforts in the recent past. As a result, its level of maturity in this area is low, indicated by point 1. In this example, the consultant enters the organization at the wrong place, resulting in a mismatch with problems for both the consultant and the organization.

8.4 Internal and External Consultants

In addition to working with external consultants, many companies work with internal consultants for specific functions. Internal consultants are usually senior employees in the organization who have experience, credibility, and a unique way of seeing both the strategic and tactical parts of the business process. They can use this knowledge to help the organization change in areas in which they can exert a positive influence. These internal consultants often have no direct supervisory responsibility. They do, however, have existing alliances. They will not typically advocate radical change that could hurt their alliances nor put themselves in positions where their level of influence is compromised. After all, major change in the status quo could disrupt their alliances, long-term working relationships, and friendships. The external consultant, who is "here today and gone tomorrow," does not share these same concerns.

Furthermore, internal consultants are not always believed at their own site. How many times have you seen a good idea suggested by someone inside the organization be dismissed, only to have the same idea resurface, be approved and implemented when suggested by someone outside the organization? Internal consultants can best influence incremental change. They can also provide support—and even some degree of leadership—for major change, but the change has to be led and owned by the site, including top management.

Internal consultants do have several distinct advantages over external ones, provided that they maintain objectivity. These include:

1. Taking a global, strategically focused view based on pre-exist ing knowledge of the site.
2. Proposing ideas to see which ones generate support. The level of support indicates the degree of site readiness for change.
3. Digging deeper because they have more time and know the systems. They also have the ability to identify real problems and real solutions.
4. Staying through the entire effort to the end, there by addressing concerns about the sustainability of change process.
5. Lowering costs relative to an external consultant.

Let's look more closely at each of these benefits.

Taking a global view. Unlike consultants who are involved tactically, internal consultants can take a strategic and a global view of the work process. All too often, the people who are supposed to be helped by the consultants are immersed in the functional day-to-day work of the company. They don't have the time to look at the strategic picture. Nor do they have the time to understand and view the changes that they want to make from a global perspective. Such a perspective, however, is of real value to an organization. Internal consultants can help develop this broader perspective. They can also help check that the projects that are developed are not in conflict with each other.

Being Ready. Projects are often not accepted when proposed. In many cases, the organization is not ready for the project. There can be many reasons for this rejection, including the possibility that the planning was flawed. Internal consultants can re-propose these projects if the organization does not accept them at first.

In the early 1990s, I was the manager for a multisite systems project. I recognized that, after the project was complete, a multisite group composed of users of the application was needed. They would work on post-project concerns such as best practices in the use of the application and how to upgrade the software as the functionality of the system improved. The idea of creating such a group was proposed, but, for many reasons, it failed to gain acceptance. Was the idea a bad one? No, it just was an idea for which the organization was not ready. I set aside the idea, and let it resurface five years later. At that time, different circumstances existed at our plants. As a result, a multisite users group was formed. This group has provided real value to the company. It has also provided benefits to the user community. The initial rejection was not because the organization didn't like the idea. Instead, the organization simply wasn't ready the first time the idea was proposed.

Digging deeper. Most consultant efforts begin with a series of interviews intended to identify the problems and recommend solutions. All too often consultants facing budget constraints, hold fewer fact-finding interviews then are needed. How often have you had a consultant ask you to send "a few key people" to the interviews? This approach has the potential to create a flawed process if the consultants extrapolate their findings based on this limited data. Evidence of this problem can be seen when the findings from their interviews don't make sense or the project leaders realize that the consultant is off track. Internal consultants often avoid this problem. Because they generally don't face the same kind of budget constraints, they can often dig deeper than external consultants can.

Staying to the end. Internal consultants are a long-term part of the organization. Therefore, they have the capability of not only proposing change projects, but also of seeing them through to the end. They can also modify them, as needed, to suit changing circumstances. Not all change processes are simply developed and rolled out. Many require course corrections. External consultants could make these corrections over time, but then they would be in place long beyond the initial project—an expensive and unrealistic approach.

Lowering cost. Quite simply, internal consultants are usually cheaper than outside consultants. Before you take this at face value, however, make sure that you factor into any internal consultant's cost all of your company overheads. This will close the cost gap, but the difference will probably still be significant.

8.5 Combining External and Internal Consultants

Do not let the previous discussion undermine the value of a good external consultant. Each type of consultant offers clearly defined value. How then do we blend them to obtain the optimum benefit? What kind of consultant alliance can we create to get the best value for the organization? If you review the four-quadrant model in Figure 8-1, you can see that Quadrant 4 represents an alliance in which the client and the external consultant share ownership and responsibility for both the effort and outcome. Figure 8-3 compares the strengths that each type of consultant offers. From there, we can see how they can be blended.

As you can see, the consultants share strengths and also offer benefits that the other can't. Your trick is to blend these skills into a multifaceted project group with both internal and external experts. This combined group will bring the maximum benefit to your organization.

You may be questioning how to blend individuals and groups that have some similar, but also some dissimilar goals. To make a multifaceted project group successful, you need a collaborative effort. As the owner, you must lead it. You cannot and must not abdicate your ownership or responsibility to the external consultant.

In particular, you must lead project design. This stage determines how you will select the consultants, what you expect of them, and how you expect them to deliver their part of the effort. All of these considerations are important. The way they are developed will dictate the type of consultant you select for the project.

8.6 How Do You Find the Right Consultant?

Finding the right consultant is not easy. However, you can follow certain steps to give yourself the best opportunity for selecting the best

External Consultant	Internal Consultant
• Broader knowledge base	• Deeper knowledge of the process
• Short-term involvement – no alliances	• Long-term involvement
• Availability of resources at many levels	• Ability to use key resources as required
• Focused effort	• Focused effort
• Better multi-industry view	• Better plant business view

Figure 8-3 Comparison Between External and Internal Consultants

consultant for your specific project. These steps follow. Modify them to suit you own needs, perhaps dropping some (not too many) or adding one of your own. In the end, you need to follow a thoughtful process if you want the right consultant.

Step #1 Recognizing that You Need a Consultant

If you are reading this book, and especially this chapter, then you have already determined that you want to consider a consultant for your effort. If you have reached this conclusion, then I salute you. You and your company have moved past the belief that you can undertake a major work process change without help. The companies that can successfully go without consultant support are few and far between.

Step #2 Finding Consultants that Can Help

Consultants are available in every conceivable form and type. Some consultants have single or multiple industry focus. Some have vast resources, whereas others are one-person companies. Some consultants not only help you with your process change, but also sell you software to support it. (These are often software development companies that have expanded into consulting to increase their presence.) With all of these consultants, where do you start?

First understand that this step is not for selecting the consultant. Indeed, it is for narrowing your choices from a great many to a manageable number. You are trying to find consultants from which you can begin a more in-depth review and selection process. You must make this first cut properly because the consultant you select will come from this group. You can develop your list in many ways. Some ideas include

1. Trade shows where you can meet and discuss your project with consultant representatives. If you follow this approach, go to trade shows for your industry or your line of work. Be sure that the consultants you see have experience in your area.
2. Discussion with others in your company who have used consultants. They can give you good leads for follow-up contacts.
3. Trade literature.
4. National, state, and local groups to which you belong.
5. Networking, especially with colleagues in your industry or in the same type of work.
6. Consultant literature you have saved. Consultants are always looking for work. When you talk to them, you may not have an immediate need for their services. Still, you should request and save their literature. Then, when you are trying to create your list of potential consultants, you will have a vast store of information.

Step #3 The Initial Meeting

Once you have created the list of consultants that you believe can help you, your next step is to meet them. You need to give them an opportunity to explain their services and what they offer. This step has two parts. The first is a presentation by the consultants describing their firm and the work they perform. These presentations usually focus on what they think your interests are based on preliminary discussions you have with them. In these initial phone discussions, the consultants will ask a lot of questions, trying to get a sense of your needs. This questioning is typical. Most consultants have varied skills; they want to be sure that when you meet, they have prepared something that fits your requirements.

Following the general presentation is a dialog in which you tell them in more detail what you want to achieve. They tell you how they can help. This dialog is valuable for both you and the consultants. The consultants get a better sense of what you want. Then if you request a formal proposal, they will be better able to develop one. From your perspective, you get a chance to see if the consultant can provide for your needs. You also get a sense of the type of work they do. You may even get ideas that, even if you don't use the consultant, you can apply to your change effort.

You must have your preliminary scope developed before these meetings. Only then can you correctly communicate your goals. If your scope is not clear, then the consultants could offer you services that seem to provide the correct approach, only you learn that it is not after they have begun. In other words, do your homework.

Step #4 The Short List

Once the preliminary group of consultants have had their opportunity to present their capabilities, you need to narrow the list further to create a short list—those firms you would actually consider hiring. To accomplish this, additional screening of their capabilities is required. Further discussion in person or by phone may be necessary to resolve unanswered questions or clarify key points about their services.

Other screening methods include reviewing their references (they will provide you a list) or having them submit samples of their work. The former is somewhat biased. Businesses generally list only satisfied customers. However, even biased references will reveal helpful information if interviewed correctly. Samples of past work can also be very helpful, enabling you to evaluate the work the consultants actually do. Most consultants will provide this information, often taking out the client's name for confidentiality.

With this added information, you and your team should eliminate consultants that do not satisfy your requirements. Your short list should now include from three to five consultants. Remember to call all of the firms, not only those on the short list, but also those that did not make the list. Let them know the results of your assessment. Let those on your short list know additional contact will be forthcoming. Be prepared to tell them what the next step will be. Once again, you must do your homework so that these next steps are clear. You should also call consultants you eliminated. Why? First, it is professional courtesy. Second, if you don't call them, they will repeatedly call you, thinking they still have a chance at the work. Third, you might work with some of these consultants in the future. Providing them feedback enables them to tailor their services to meet your needs.

Step #5 Detailed Scope and Budget

At the same time you develop your short list, you need to develop the detailed scope of the project and the associated budget. No doubt you already have a very good idea of what you want to accomplish, especially if you have worked through the vision and Goal Achievement Model. However, to build a good working relationship with the consultants, you need to be specific and very clear on what you expect them to do and what you expect them to deliver. The scope of work needs to be prepared by the project team so that, when completed, you are sure that it addresses all issues. Furthermore, senior management needs to review and approve it. This not only achieves buy in, but also allows you to use the document as a scope control tool later in the project.

Also included in every scope is the cost. This is the budget for the project and needs close monitoring and control. Lack of control, and its associated cost overruns, can spell failure for even the best effort. The cost needs to be included in the scope along with a cost-benefit analysis that shows the value you are getting for the money. Without a payback on the investment, the project probably will not proceed.

STEP #6 THE SECOND PRESENTATION

Step 4 and Step 5 need to proceed simultaneously. At the completion of these two steps, and after the unsuccessful consultants have been notified, you need a second presentation from those consultants still under consideration. This meeting is very different from the first "get acquainted" meeting. At this point, you are addressing the specifics of the job and the approach that the consultant plans to take. To make this meeting successful – and lead you to your final selection – you need to provide the consultants with the detailed scope of work. They will need this information in advance of the meeting to allow them time to fully prepare their proposal and associated presentation.

STEP #7 FINAL SELECTION

After the consultants on the short list have made their presentations and have responded to any questions you have, it is time to make your final selection. Ideally you have been part of a team making this decision. If not, and the selection has been left solely in your hands, you will have the full responsibility if either the effort or the consultants fail.

The final selection is often difficult, especially when more than one final candidate is fully qualified and within your budget. The decision is also a somewhat subjective one at times. The more subjective it is, the harder it is to explain your rationale to senior management. For these reasons, make the selection process as objective as possible. For example, develop a set of weighted evaluation criteria with an associated scoring process. Each team member can then independently evaluate the candidates and provide an objective score. These scores are then combined to create a team ranking. Some examples of evaluation criteria include: experience working in your industry, experience working in your specific function, number of consultants currently available, resumes of consultants who will work on the project, project references, and samples of their work.

The following steps describe this process:

1. Develop the evaluation criteria.
2. Assign a value to each criteria –the maximum number of points a consultant can score for that criteria. The values do not do not all have to be the same. Some criteria may be more important to you than others.
3. Provide the score sheet to the evaluation team before the meet ing with the consultant
4. Make sure everyone on the team understands the criteria and how to conduct the evaluation.
5. As soon as possible after the presentation, each member should independently evaluate the consultant providing a score for each criteria up to the maximum value assigned.
6. Each evaluator should provide written comments to support their scores.
7. Combine the scores for a final team score.
8. Repeat for each candidate.

When the entire process is completed, the team can review the scores and select which consultant will be awarded the project. As with the short list process, call all consultants who participated in the final selection process thanking them and advising them of the results.

Step #8 The Contract

Once you have made your final selection, there is always the contract phase. Here, qualified people within your company finalize the terms and conditions of the working relationship in the form of a legally binding contract. This often requires reaching agreement on terms and conditions developed by both companies, which may or may not be compatible. Leave this part of the process to the experts, but stay involved. You must remember that once the contract is approved it becomes the binding agreement for how the companies will work together and resolve issues. Therefore, in the best interests of the project, you need to fully understand the agreement.

Step #9 Getting a Feel for Your New Partner

After the consultant has been chosen, some companies still want to proceed slowly, making sure that the decision and the direction for the project are the correct ones. One way to accomplish this is with a pilot program. Setting up a pilot project—a smaller version of your full-scale effort—isn't always possible. However, if you can set one up, there are several benefits. You will learn more about the consultant's level of experience and the working relationship you can expect to have in the larger scale project. You can also test your process design on a smaller scale. A further benefit is that the consultant can get to better know you, your requirements, and the effort in general.

The pilot approach isn't always possible. However, there are other ways to build your relationship with the consultant. One way is to divide your change effort into several stages. Then contract with the consultant one stage at a time. This approach gives you the opportunity to evaluate the experience and working relationship. You can then make appropriate adjustments. This approach also gives you a way to terminate the effort at various break points in case the consultant is not working out the way you wanted.

8.7 How Clients Can Hinder the Process

Let's assume that you decide to use an outside consultant for your change project, and have gone through the selection process. The consultant is now starting work. Another assumption has been made automatically: that the consultant can help. Under most circumstances, this is true. However, even the right consultant doesn't always help. The risk of failure—or a false start in a change effort—can be extremely dangerous to the organization. Efforts of this type that fail can seldom be restarted until years later. Most organizations simply cannot deal with change efforts too often.

Even if the project has been properly developed and the consultant has been carefully selected, what can get in the way of success? There are many factors including the following.

1. You expect too much too soon.
2. You do not provide sufficient information or communication.
3. You do not provide visible support to either the consultant or the process.
4. You do not have a vision, a goal, or a detailed roadmap for the effort.
5. You overcontrol or undercontrol the work.
6. You abdicate control or ownership.
7. You fail to provide for the needs of the organization, consultant, or the process.
8. You do not adapt as the changes evolves. Recall the discussion about spiral learning.
9. You over extend or under extend yourself.
10. You create adversarial relations.

As you were reading this list, you probably noticed that all of these factors were problems that the client caused. Can't the consultants cause problems as well? Yes, consultants can, as you will soon see. However, they have experience in the change process; it is after all their business. The client on the other hand, has less experience and is often the prime contributor. There is no easy answer to this problem. You must pay close and ongoing attention to the effort as it evolves, identify problems that you are creating early, and correct them.

8.8 How Consultants Can Hinder the Process

In many cases, the consultants are the ones to cause a project to fail. Let's look at some of the ways that this can happen. Given this information, you can look for the symptoms and avert problems that could occur from lack of attention.

The consultants deviate from the original scope. Some consultants are trying to generate more work or revenue for their firm. By wary of consultants who constantly suggest additional work beyond the scope for which they were contracted. Don't confuse this problem with cases where the consultant legitimately identifies of work that may be out of the original scope, but is now needed.

If you and your consultant are operating from a Quadrant 4 position, then you should not have surprises of this sort. The additional scope would be recognized and addressed by the team. You can also avoid this problem by having a rigid scope control process. Then your response is that while it would be nice to add whatever the consultant has suggested, the team needs to address the suggestion and the additional cost before approving it. The additional work can always be done later.

The Consultants are overcontrolling of the project. Remembering Figure 8-1, overcontrol is Quadrant 3 behavior—the consultant's level of control is high and yours is low. If the consultant is pushing you into this quadrant, even though you initially started out as partners in Quadrant 4, then you need to have a serious heart-to-heart discussion with them and correct the problem. As you enter into the discussion, however, keep an open mind. The fault may not actually be their responsibility. It can also develop if your team has abdicated responsibility.

The consultants have the wrong people on the job. This problem is a very real possibility. The consultant may have applied enough resources, but the wrong resources to the work. They may have misunderstood your requirements. In addition, the people they intended to assign to your project may have left the company or be tied up longer than expected with another client.

They may also be trying to work through a cost issue without upsetting you. This last problem is something you need to watch for, especially if you have pressured the consultant to spend less money. The level of resources they provide is based on your scope and what you are willing to pay. If you have a set budget and want specific resources, they will attempt to accommodate you. You may, however, be assigned resources that are not the right ones for your work effort. If this occurs, your team needs to review the requirements established at the beginning. Did you expect more than you paid for or has the consultant delivered less than agreed?

The resources that the consultants provide lack the required experience or have the wrong experience. If you get the wrong resources due to cost issues, or if the consultant sends the wrong resources, you need to correct this problem.

The consultants have the wrong philosophy about how to run the project. If you have done your homework and properly selected the consultants, this problem should never happen. At minimum, you should have the right consultant based on your preparatory work. If, however, you didn't do the work described in the section on consultant selection, then shame on you. If cutting corners at your end resulted in hiring the wrong consultant, you need to correct the problem sooner rather than later, if you still want your project to have any opportunity for success

8.9 Working as Partners

Change management consultants can help your effort be successful. However, even with all that consultants bring to the effort, you must not abdicate your ownership of either the effort or the long-term outcome. You are there for the long haul. The external consultants are gone when the contract is over. Look at the consultants as partners, but you own a majority of the stock. If you approach your change effort in this manner, you will be successful.

Chapter 9

Resistance to Change

9.1 What Is Resistance to Change?

By now, you should have a good understanding of what change is. The question remains: Why doesn't change always happen? You may start with a good idea. But somewhere between the creation of the good idea and its the actual implementation the process gets interrupted. The change, as good as it is, fails. In this chapter, we look at why this happens and what you can do to keep change moving forward.

Change often fails due to bad planning or bad execution. But what if you have taken the time to carefully plan the change effort and have worked hard to execute it? The process you put in place still can flounder. It may never reach its full value or, worse, fails entirely. The critical factor may be resistance to change, a condition present in every change effort and in every work environment. In spite of your good ideas, many people will see them in a different light. They believe that the change is not in their own best interest or the company's. They then act on this belief, doing whatever they can to stop the change from happening. At the least, they try to make the change as minimal as possible.

Chapter 9

Some people who propose change have the attitude that their position of power, or even the simple obvious value of their idea, will outweigh those resisting, so that in the end the change will take place. This approach is extremely dangerous. Resistance, if ignored, will ultimately destroy even the best idea. If, however, you can take preemptive action, then you can often minimize the problem and, perhaps even make it go away.

9.2 Why Do People Resist?

Why do people resist change that is clearly to their benefit? The simple answer is that they do not view the change as an improvement. Often, if asked, they will tell you they see it as a step in the wrong direction.

Many people resist because they sense a mismatch between the new environment (what you are trying to do) and their comfort zone—the area in which they operate on a daily basis. Within this comfort zone, they do not feel threatened either by the work or the environment. This state is often called the status quo. Take people out of their comfort zone and they not only feel uncomfortable, but they also do whatever they can to restabilize their environment. Sometimes this is easy for them to do. But at other times, protecting their environment can be difficult or outright impossible. At these times, an individual's level of stress increases and they try even harder to restore the status quo. Thus, a critical component of this comfort zone model is that the further you take people beyond this zone, the more that stress levels increase to the point where they become unbearable. The stress can become so severe that they must try either to restore the status quo or to expand or shift the area of the comfort zone to include the new set of conditions. In this way, even though you did not restore the status quo, you adapted comfort zone to include the new change. In both cases stress is lowered and comfort is restored.

You probably have had your own experiences when you have felt outside your comfort zone: a new job, a major change in your job assignment, a layoff, or even the many events that occur regularly in your personal life. In each case, you have either tried to restore the old way or tried to embrace the new. In all cases, however, what you were trying to do was to restore a comfort zone that would let you continue to operate stress free.

When you can control change, you have a greater likelihood of returning to your previous state. A simple example involves your car. If you are like me, you need your car to get to and from work each day. But what happens when the car breaks down? You need another way to transport yourself. This may involve borrowing a car or getting a ride from a coworker. You have a change to your status quo and, as a result, your pattern of work is changed. You may feel uncomfortable because you are driving a borrowed car or have to be dependent on someone else for a ride. You are out of your comfort zone, and, while you may not refer to it as such, you feel discomfort. In most cases, though, you have the car repaired, thereby resuming the status quo and returning to your comfort zone. In this example, the car repair is a change you can control. With relatively little money and time from your mechanic, the status quo is restored, stress is reduced, and your world returns to normal.

Now suppose your car does not just have a few problems, but breaks down beyond repair. Here it is impossible to return to the status quo. What do you do? Because you need a car, you go and buy one. Your comfort zone is altered in several ways. First, you now have a new car that you may not want. Second, you probably have debt (the car loan or lease payment) which you also do not want. When you went through the process of buying the car and taking the loan, you probably felt discomfort because you were operating out of your comfort zone.

Now think about what happens when you drive your new car off of the lot and the car-buying process is behind you. You may feel pretty good. You have a new car and don't have to worry about it breaking down for a long time. Why did this transition of feelings happen? You encountered a problem that took you out of your comfort zone. You could not restore the status quo. Therefore, you mentally changed the boundaries to include the new event. Once you made this shift, the new car and loan were back in your comfort zone.

Now let's consider a work process redesign effort. What is different about this change compared to the simple individual changes just discussed? This broader change affects not just you, but an entire organization. Even if everyone is agreeable, expanding the collective comfort zone is not done easily or quickly. The stress level that results is high, having a negative impact on those involved. The transition could be made easier with planning and sound execution. However, the issue of the comfort zone still remains.

Faced with change, what does the organization do? Assume that planning and execution are not an issue. The organization has two choices. The first choice is to collectively expand its comfort zone to encompass the new set of processes. The second choice is to try to reduce or even eliminate the new set of processes, thereby allowing a return to the former state. Either of these choices works. The former, however, can lead to progress (assuming the change is for the better), whereas the latter can lead to stagnation.

Suppose you have a change that you cannot control, nor can you easily expand the comfort zone to fit the new set of circumstances. For example, you may face a major restructuring at work, one in which many of your tasks are eliminated, jobs changed or eliminated, pay grades reduced, and other equally disruptive actions. A change of this magnitude is serious. The collective comfort zone of the group is not easily restored and people will need a long time to readjust. Furthermore, if this transition is not well thought out, resistance will most certainly emerge.

Let us consider two more examples. Suppose that your plant is implementing a new computer system. In your opinion and that of many co-workers, the current system works better than the proposed system. You don't want the new system, but it has been mandated by corporate management. As you learn the new system, you discover major problems with its report writing function, providing clear cut evidence to back your concern that the new system is a step in the wrong direction. Fortunately, the project team is receptive to input from your plant, and listens to your concern about the report writing function. In fact, the project team specifically addresses your concerns. It adapts the new system so that it delivers reports with the same level of detail you had before. You still may not like the new system, but now you feel better about it. Why? Because your specific concerns have been addressed and you have been able to readjust the boundaries of your comfort zone to accommodate your needs.

Now suppose your plant decides to put a work process redesign in place. It is badly planned and badly executed. The expectations for you and your co-workers are not at all reasonable. You note specific problems to the project team, which ignores you. The process is implemented and no way is available to return to the previous mode of doing business. Most of your co-workers are frustrated and several are quite angry as a result of their comfort zone being destroyed. Open resistance to change develops. The employees comply only on a superficial level, hoping that—like many previous management initiatives—this one is a "program de jour" that will eventually go away. The organizational stress goes to a level that can not easily be mitigated. The organization's effectiveness and efficiency plummets, resulting in lost production and profit. Resistance can indeed be costly.

9.3 Forms of Resistance

Resistance takes on many forms. Before you can overcome resistance and have a successful work process change, you need to be able to recognize them. Some forms of resistance are obvious and easy to recognize. Others are very subtle; if you are not paying attention, they will undermine or even destroy the change effort before you can react.

Resistance can be categorized in four forms, depending whether the resistance is active or passive and whether it is open or hidden. Figure 9-1 illustrates these four forms.

The *y*-axis measures the visibility of the resistance by those affected by the change. This action can be open; the resistance to the new process is obvious for all to see. The action can also be hidden. In this case, the resistance is below the surface, not easily seen. If you are paying close enough attention, you may notice it, but it is more difficult to identify.

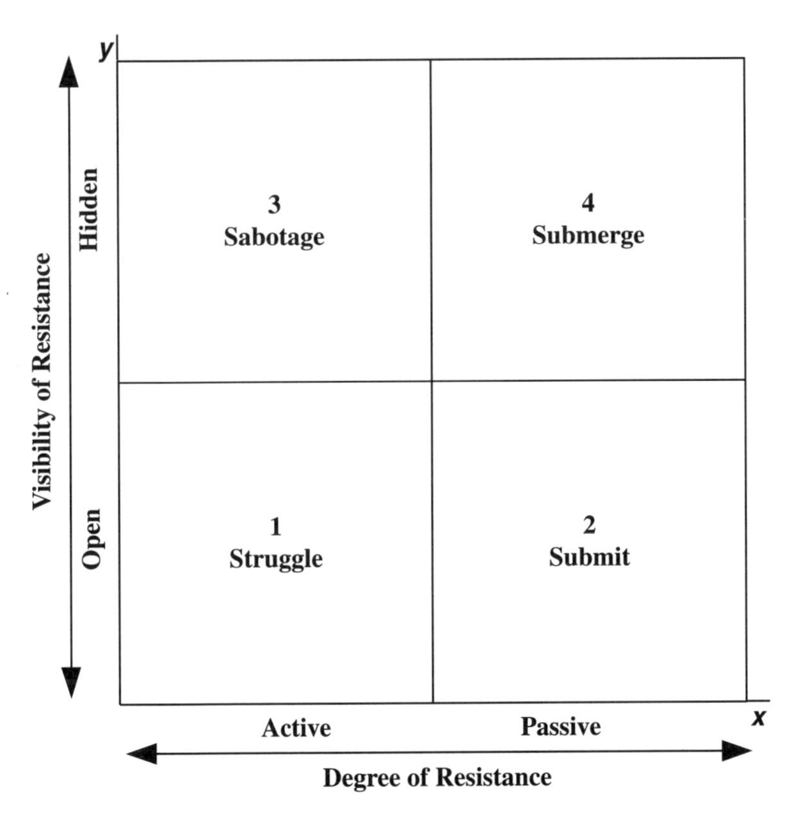

Figure 9-1 The Four Forms of Resistance

The x-axis identifies the degree of resistance, whether it is active or passive. Active resistance is designed to stop or hamper the change process. It can be destructive to the organization because, as the resistance evolves, people may be forced to take sides, leading to substantial business problems. Passive resistance is more difficult to confront because it is generally less obvious. Rather than trying to actively block change, people engaged in passive resistance are not going along with the effort. They might continue to use the old system, ignoring change for as long as they can. They may work more slowly, take extra time off, and keep "forgetting" how to incorporate the change into the work process.

As Figure 9-1 illustrates, the resistance model has four quadrants. Each is discussed below.

Quadrant 1—Active, Open Resistance—Struggle. You begin to implement change when some of the employees openly tell you the change is wrong or, even worse, they won't go along with it. A scenario like this one should never happen. Change should be discussed and buy in achieved before you try to implement it. If you have planned the process change and the subsequent implementation, your organization should never have to be placed in this situation. Otherwise, you can expect active and open resistance. The one good aspect of open resistance is that it indicates that your employees feel comfortable enough with you as the manager to tell you openly how they feel about what you are doing. If they don't trust you, the resistance will be hidden leading to Quadrant 3 behavior.

Quadrant 2—Passive, Open Resistance—Submit. Passive resistance is quite different from active resistance. It involves people submitting to the new order of things, in a sense "going along." Don't mistake this submission for acceptance and open embrace of the change. Even though they do what's necessary to make the change work, you lose their energy, enthusiasm and loyalty. Unless you win them back, you may see a gradual downturn in productivity and increased turnover.

Quadrant 3—Active, Hidden Resistance—Sabotage. If your employees do not trust you, resistance to change will take on a different form. In the Quadrant 1, employees didn't feel they were threatening their own security if they told you exactly how they felt about the changes you were making. In Quadrant 3, they do feel threatened. They will resist change as actively as the employees in Quadrant 1, but will try to hide their resistance, sabotaging your efforts. At least in Quadrant 1, you knew what resistance you faced and could respond to it. Not here. The problem often comes from management style. Employees want to be trusted. If, however, the management style at your company is that employees should simply do what they are told, then resistance will be hidden and you will face sabotage.

Quadrant 4—Passive, Hidden Resistance—Submerge. Unlike Quadrant 2 resistance, Quadrant 4 resistance is hidden. Because resistance is passive, it frequently is not as severe as sabotage (Quadrant 3). Nevertheless, it is still dangerous. Your employees are indirectly saying that they will do what's asked, but will undermine the effort at every opportunity. At least with active, hidden resistance you are aware of the resistance once it has occurred. With Quadrant 4, you can not see it, it is submerged. On the surface, everything may seem fine. Meanwhile, below the surface, you face severe problems. Quadrant 4 resistance is used to undermine change. Your process may fail and you may never know why. Nor will you have anyone specific to blame for the failure, except maybe yourself.

9.4 Overcoming Resistance

Once you know about resistance and can recognize it, the question becomes, "How can you overcome it?" Sometimes you simply can not. However, even when you can not make it go away, you can minimize it, essentially making it a nonevent.

Let's look further from a business perspective at how resistance emerges. People generally resist when they move out of their comfort zone. The basic cause is often the lack of agreement between the goals that you are trying to achieve and the goals of the individual or group with whom you are trying to achieve them. If the goals, or change initiatives, are in agreement with the employees' goals, you have no problem; the comfort zone is not being violated. The comfort zone is expanding with the agreement of the employees who have to adjust to its growth.

The problems begin when your goals are not aligned with the employees' goals. The key then is to put change initiatives into place in a way that allows the goals of the two forces (you and the employees) to be aligned. With alignment comes success. Your success is shaped by two important factors. The first factor measures how well the goals agree between those trying to implement them and those having to do the work. The second factor measures the balance of power between the groups. These factors dictate the approach you need to implement the change process successfully.

Figure 9-2 provides a four quadrant diagram that illustrates different ways change is implemented. In this figure, the x-axis measures how well your goals are in agreement with those of the employees who have to implement the change. The y-axis measures from high to low the balance of power between those who are trying to implement a change initiative and those who are responsible to make it work.

The balance of power, in conjunction with the amount of agreement on goals, determines the level of resistance to the change initiative. A high balance of power corresponds to a team relationship. Both sides work

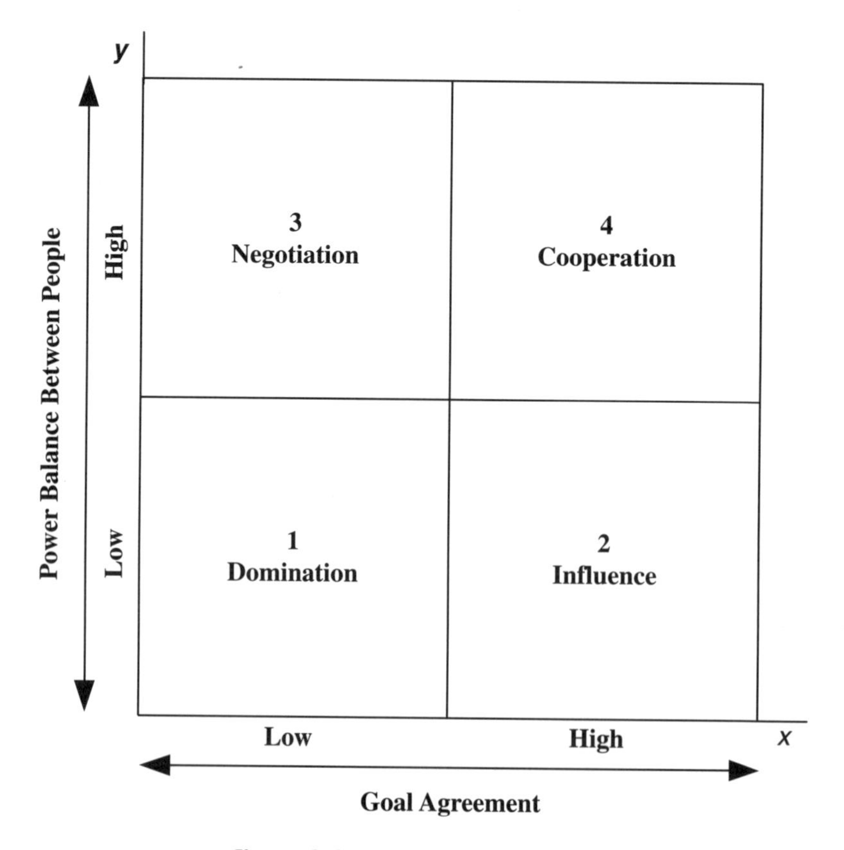

Figure 9-2 Implementing Change

together to accomplish the change efforts. A low balance of power indicates that you and the management team will dictate to the employees what will and what will not be done. This relationship could easily be described as "command and control." In addition, the balance of power influences the way in which the change initiative is presented to the organization. It affects the type of resistance that may emerge. The balance of power can also help you determine the best way to rollout the initiative so that you minimize conflicting goals and resistance.

Quadrant 1—Low Goal Agreement, Low Power Balance—Domination. In Quadrant 1, you and the employees share no agreement about the goals of the organization. In addition, the power balance is low; when goals are presented, they are usually in the form of "here is what you are going to do." The combination of low agreement and low power balance creates an environment in which you dominate the employees. In a domination model, resistance usually takes the form of sabotage (if employees resist actively) or submergence (if employees resist passively). You tend to see submergence more often. Those who openly resist generally no longer work for the company, whether by their choice or by management's. The evidence of submergence is that after dictating the goals, management repeatedly wonders why change is never implemented or, if it is, it is superficial at best. When the organization acts in this manner, it usually doesn't recognize that its entire approach is wrong, let alone understand why its initiatives are failing. The best way to avoid resistance in this quadrant is simply not to be in this quadrant. In today's business environment, the best managers engage the employees, rather than lose all of the value that they could add.

Quadrant 2—High Goal Agreement, Low Power Balance—Influence. In Quadrant 2, you and the employees agree on the goals; you are in alignment. With this relationship as a staring point, change is much easier to implement because there is no conflict about what needs to be accomplished. The only downside is that the power balance is low. This imbalance doesn't usually lead to a great amount of resistance, but you still may encounter some. Because the power balance is low, you must use influence to get your initiatives implemented. Although the employees agree with the goals, many problems can occur, based on how you apply influence. If you take advantage of the power imbalance and try to dominate the employees, even those who agree with the goals will resist your efforts. Influence can be applied in many ways: personal charisma, rewards, and cheerleading among them. Especially because the power is on your side, employees will be encouraged if you take time to listen to their views. You need to find what works best for you. One easy method is to engage those who you are trying to influence. Gather a group of key personnel, discuss what you are trying to achieve, and use their collective input to determine how to best influence the work group.

Quadrant 3—Low Goal Agreement. High Power Balance—Negotiation. In Quadrant 3, the balance of power is shared. Therefore, you and the employees are in a good position to work together to implement the change effort. Unfortunately, the agreement on the goals is not strong, leading to the potential for resistance. Your best strategy, based on high balance of power is of negotiation. In many ways, your strategy here is similar to your Quadrant 2 strategy. However, the methods of reaching goal agreement are more sophisticated. In fact, it is through the negotiation process that goal agreement will be reached.

How you manage the negotiation process will determine the kind and degree of resistance. As you may already know, there are many types of negotiations. The traditional model still used in too many union-management negotiation processes is the win-lose model. Each side puts its issues on the table. Then through a process of withdrawing or conceding items, the two sides eventually (we hope) arrive at something acceptable to all. To get to this point of agreement, much is often lost. What the employees must concede—how important the items are—influences their level of resistance. The better approach is to conduct negotiations in a manner that leads to win-win result in which both sides are happy. Many books specifically about negotiating are available on the market. When you conduct a win-win negotiation, the focus should be on what's good for the business. Conducting a win-win negotiation is not easy. You may want to have an experienced facilitator present.

Quadrant 4—High Goal Agreement, High Power Balance—Cooperation. Quadrant 4 is relatively easy. In this quadrant, you and the employees share a balance of power and agree on the goals. The change initiative should be successful. Furthermore, you should encounter no resistance to implementing it.

9.5 Helps and Hindrances

Suppose you have the best case going for you. You are in Quadrant 4 with full goal agreement and a shared balance of power – everyone is willing to work together to make the change happen. This is still not enough. Certainly resistance will be minimized; you are in a good position. However, you must consider additional factors. The main one is how you approach the change. How do you expand the collective comfort zone so that people feel comfortable with the changed environment? Some of the ways for growing the comfort zone include the following strategies:

Grow Slow. Recognize that change does not happen overnight. Therefore be patient and grow slowly. Before you get impatient, think about how long you took to get where you are. This may have a sobering effect and promote the patience you need. As a good friend once reminded me, in the world of change, "slow is actually fast."

Grow with People in the Know. Most employees will be more than willing to give the change an opportunity to succeed if you have taken the time to talk with them. Let them know not only what you are doing, but also why you are doing it. People want to have a level of understanding when their world changes. Recall how you felt as a child, when adults explained change with the infamous words, "because I said so." You probably disliked that expression very much. The same holds true for those who are affected by your change initiative.

Grow with Involvement. Having employees understand what's going on is only part of the successfully implementing change. For employees to add value to the process, they must be allowed to participate. Allow, even encourage, those who will be affected by a change to help design how the change will take place. They will be more willing to embrace the change. As a result, resistance will be minimal.

Grow with a Plan. Going slow, communicating, and involving those impacted are all important. The most important factor for success, however, is having a well-thought out plan for how your effort is to proceed.

Address Resistance. No matter what you do, some amount of resistance will occur. The easiest way to overcome it is to address it head on. In this way, you can keep it from growing and negatively affecting the positive changes you are trying to bring to your organization.

Chapter 10
The Web of Change

10.1 What is the Web of Change?

Chapter 7 introduced the eight elements of managing change: leadership, work process, structure, group learning, technology, communication, interrelationships, and rewards. Individually, each of these elements is important and will be discussed more fully in the following chapters. However, their combined importance far exceeds the sum of all of the individual elements. If you change one of the elements in the set, even slightly, it will effect some, if not all, of the others. This chapter introduces a tool—the Web of Change—that brings the eight elements together, describes their relationship, measures how it changes over time, and recognizes the impact that each element has on the others. Figure 10-1 illustrates a sample web. To accomplish this I have developed a diagram that I refer to as the Web of Change.

Figure 10-1 is a radar diagram. It enables us to view all of the elements on the same chart at the same time. A radar diagram is used when you want to show data points for multiple variables on a single diagram. It is called a radar diagram because it looks like a radarscope used by air traffic controllers. It is created as a series of concentric circles. Each circle represents increasing values of the variables.

Chapter 10

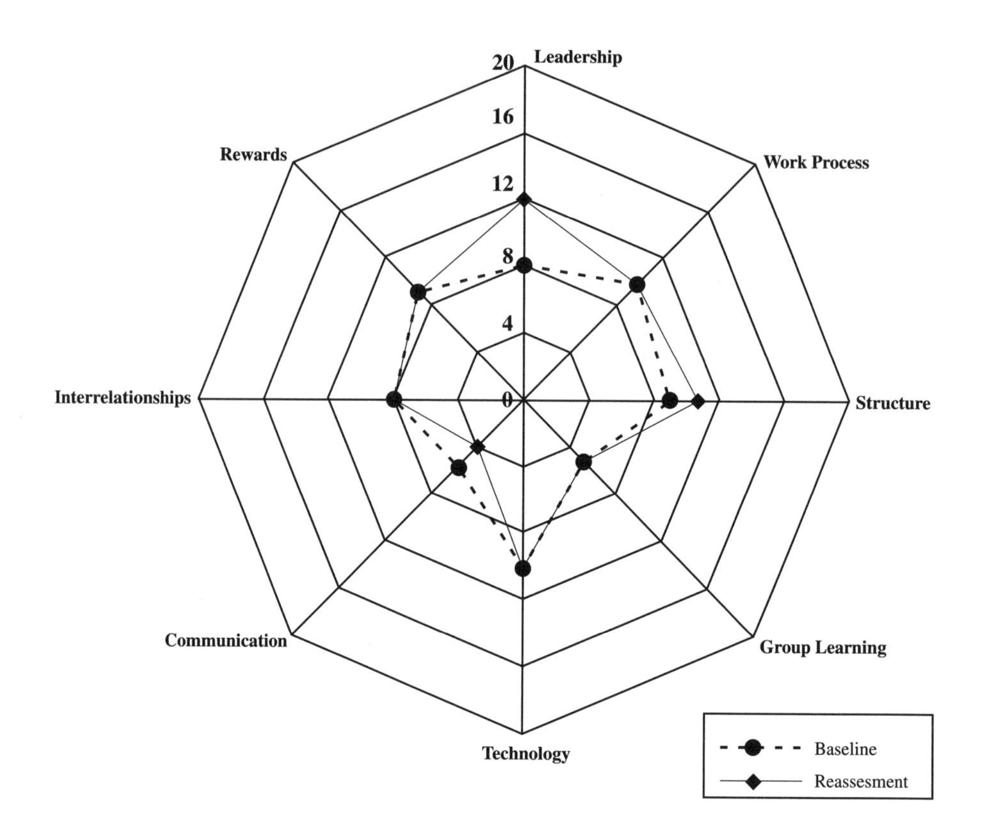

Figure 10-1 The Web of Change

The diagram in Figure 10-1 has five concentric circles valued at 4, 8, 12, 16 and 20, increasing in value as you move outward. Radiating from the center of the diagram are straight lines, each one representing one of the variables. Figure 10-1 has eight radial lines corresponding to the eight elements of the change process. Points of data for each of the variables can be shown on a single chart so that you can see how they interrelate. However, you can only show a single data point for each of the variables on any given diagram with a given set of data.

As you can see from Figure 10-1 the Web of Change resembles a spider web. The radial spokes hold it together. A change in any one part of the web has a ripple effect on the other parts within the structure.

Each spoke is a separate element of the change process. The web must be in overall balance to function properly. A movement or alteration in any one of the spokes or data points alters the balance; some form of responsive action, positive or negative, is required. If a few of the radial strands are weakened or damaged, and corrective action is not taken to repair the damage, the web will be destroyed. While the other strands may still be individually strong, its structural integrity is gone.

Like the spider, those of us who manage change operate within a web, in this case a change management web. We detect any movement or alteration in the condition of our environment. We must respond to any changes in any of the structural spokes in the web. Responses can range from simple rebuilding of a damaged strand to fighting off more serious and overall damage. The trick is to detect these changes so that we can respond. The web is a measurement tool that detects changes over time and changes in the work environment.

This chapter describes the Web of Change in greater detail, including how to make your own web. It identifies the elements that constitute the radial spokes and explains why they are located in their positions. Chapters 11 and 12 then describe each of the elements in greater detail. You will see how the web can be used to show the current state of change within a company. You will also see the impact that the elements have on each other and on the system as a whole.

10.2 Constructing A Web

You can easily create your own web using a computer and any of the various drawing programs available on the market. For my purposes, I have found VISIO to be a drawing tool that makes creating and modifying the web a very simple task. This web is also very easy to create because the spokes representing the eight elements can be set 45 degrees apart. Figure 10-2 shows the basic web diagram with just the concentric circles and the radial spokes. The spokes are 45 degrees apart and the web has five concentric circular rings. Eventually you want to use the web as a measurement tool; the rings correspond to values of 4, 8, 12, 16 and 20 points.

The next step, shown in Figure 10-3, is to add the elements back into the diagram. The elements are not placed randomly in their positions. Each one relates to those that follow, going in a clockwise direction.

Leadership (the most important) drives the **work process** which, in turn has a great deal of impact on the **structure.** The structure determines the ease with which **group learning** can take place.

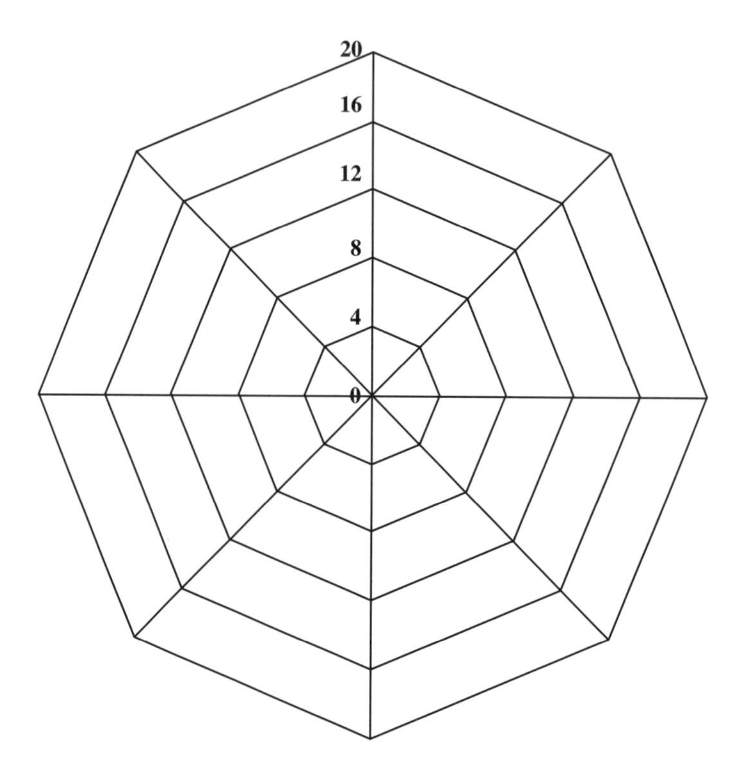

Figure 10-2 The Basic Web

Meanwhile **technology** provides tools that support all of the previous elements. As the change begins to take place, improved **communications** is the result. This drives improved **interrelationships** and **rewards** for group performance.

10.3 What Is the Web's Use ?

The Web of Change has two major functions. First it illustrates how the entire set of elements changes over time from the starting point, or baseline of the effort. Second, it shows how the individual elements relate to each other. Both of these functions are equally important.

Whether you are just starting your change effort or have been engaged in it for a while, it is important to track if change is actually taking place. Conventional metrics may tell you if you are becoming more profitable, reducing backlogs, maintaining smaller inventory, or achieving other similar goals. These metrics, however, do not directly portray how

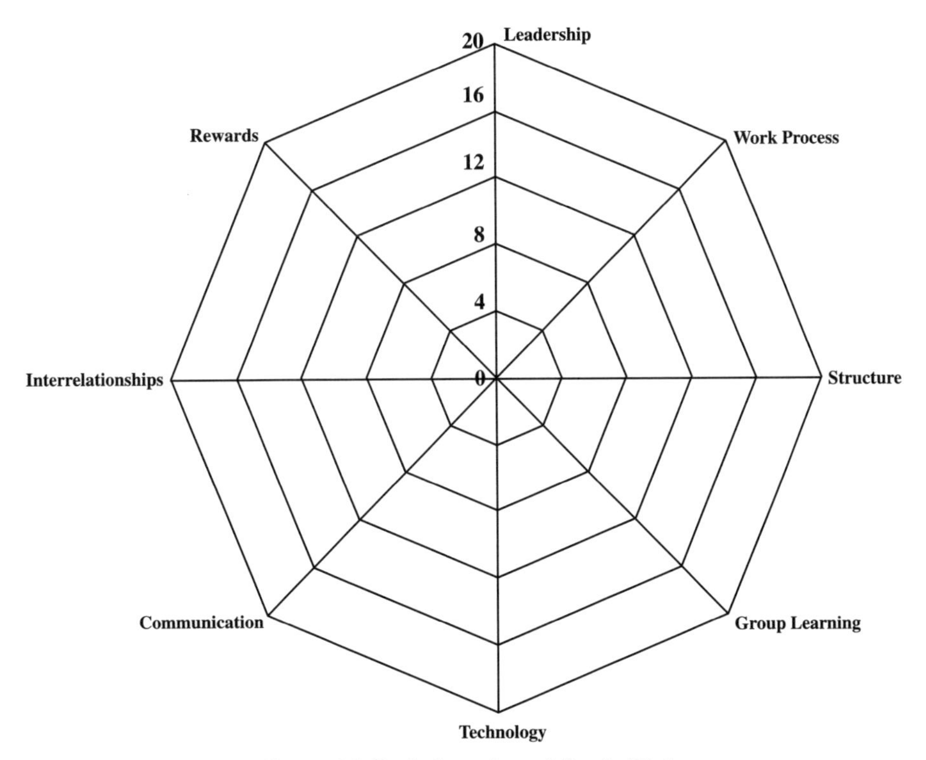

Figure 10-3 A Completed Basic Web

your efforts to introduce change are progressing. The Web of Change enables you to measure the change elements at a starting point in time—the baseline. You can compare this baseline set of data points with another set of data points at a later date, measuring the changes that have taken place. Evaluating the differences provides important information about the success of your efforts. This evaluation will also point out problem areas. For example, it may show no change in areas where you had expected some, or negative change when you expected positive results. This valuable information permits you to take proactive steps to improve the effort.

The web also enables you to measure the effect a change in one element may have on the other elements. This type of analysis tells you if something you have implemented in one area has had a negative impact on other areas. An example of this is shown in Section 10.6.

10.4 Where Do We Get the Data for the Web ?

Chapters 11 and 12 describe in further detail each of the elements for managing change. The appendix and a disk enclosed with this book, both include an eight-part survey based on these elements. Complete the survey, then develop your score according to the directions provided. These scores for the individual elements—ranging from zero (0) to twenty (20) points—become the data points on your web. The survey will produce different answers depending on where your firm is in the change process. You will complete the survey several times during the change process. Your scores will change, not only for the individual element under consideration, but also for other elements that are impacted.

As you complete the survey, you generate eight scores (one for each element) ranging from zero (0) to twenty (20) points. You can draw your web diagram freehand, using Figure 10-3 as a guide, or you can use the software provided with this book to automatically chart it as seen in Figure 10-4. The result from the first time you complete the survey will be your

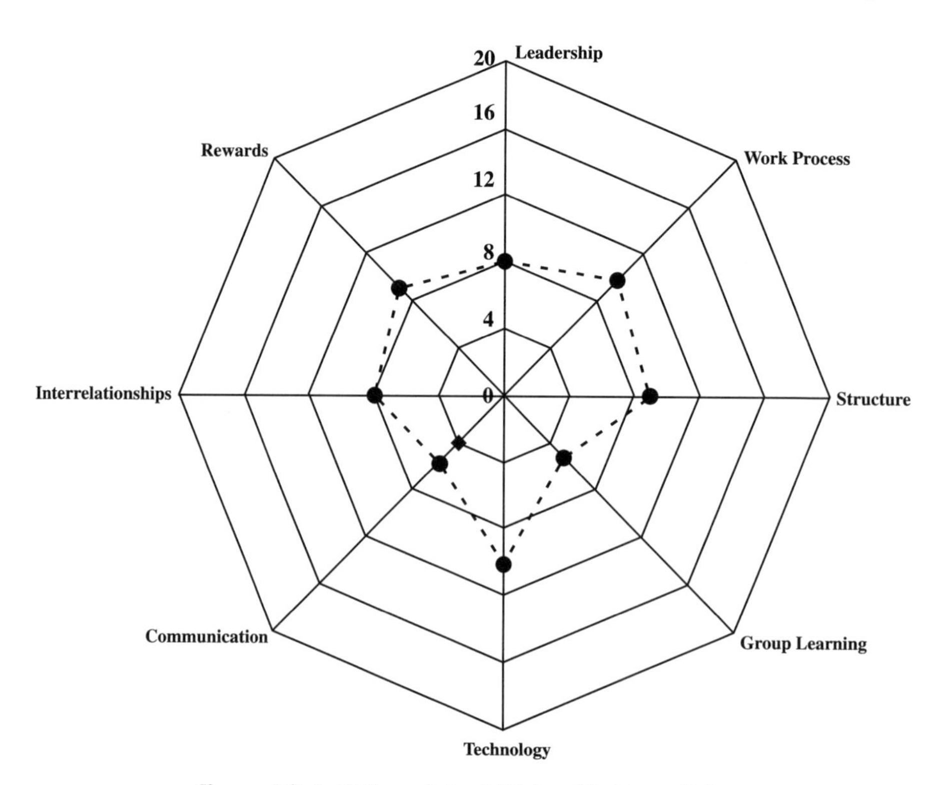

Figure 10-4 A Completed Web with Data Points

baseline. For subsequent surveys, you will be able to see the changes that have taken place over time and how the changes in different areas have impacted each other.

10.5 How Often Should We Measure ?

The question often arises about how often to measure the elements. Many people relate the change process to projects on which they have worked. However, these projects were usually of relatively short duration; as a result, the measures changed rapidly. The process of major change is the opposite. A long time and a lot of effort are needed to make change happen. If the measurement frequency is too short, no significant change will appear on the chart. In this case, the measure will be of little value.

How often is enough should be determined by how rapidly the process is proceeding and whether or not you believe there is value in measuring at a given time. Real change is long-term, never-ending process of organizational evolution. In most cases, measuring on a quarterly basis will be appropriate. Measuring less frequently will not give you enough advanced warning of problems, represented by movements in the web. Measuring more frequently will not give enough time for change to take place.

Realize too that surveys of this sort are subjective. They are based on opinion, not on hard facts. In order to maintain consistency, those who conducted the baseline survey should, if possible, be the ones who take it again. Following this approach increases the consistency of the information.

10.6 An Example of Change over Time

Let's now look at an example of the web in action. Assume that your company has decided to institute a change process, moving toward more of a team structure than the current more autocratic one. In order to create your baseline, a cross section of key employees answer the survey found in the appendix. The composite results of the survey are shown in Figure 10-5. The comments in Figure 10-5 offer a possible explanation for the scores shown in this table. Figure 10-6 then shows the initial web that can be created from the score

This web is the company's baseline for its change process. Notice that the scores for each element are represented by a point on its corre-

Element	Score	Comments
Leadership	8	The organization is strictly hierarchical with a clear chain of command structure.
Work Process	10	The work process is cumbersome due to poor inter departmental linkages
Structure	9	Strict hierarchy
Group Learning	5	Minimal learning. No feedback processes
Technology	10	Adequate
Communication	6	Only as set up by the hierarchy
Interrelationships	8	Poor across departmental boundaries
Rewards	9	Individual performance rewards. None for teams or groups

Figure 10-5 Sample Baseline Scores

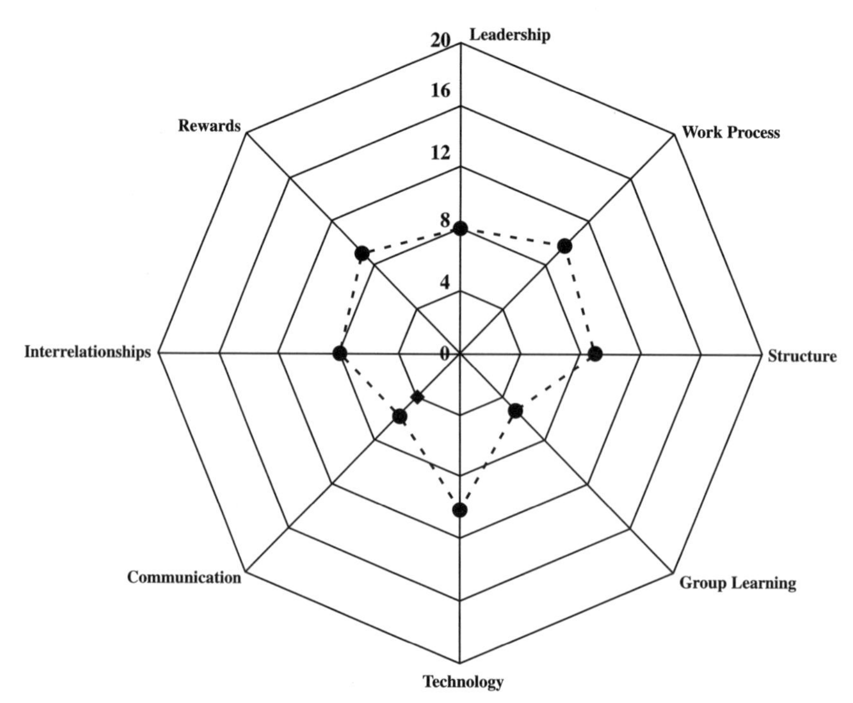

Figure 10-6 The Web of Change with Baseline Scores

sponding spoke. As the web changes, these initial data points will be kept in place, while a new set of points based on a subsequent survey will indicate the changes that have taken place.

Suppose that three months have elapsed. The group resurveys itself and conditions have indeed changed within the hypothetical company. The manager—who had imposed the military-like chain of command structure and reporting relationships—has left the company. The new management has started to build teams, vastly improving the way that the work system operates. The new survey scores are shown in Figure 10-7. In turn the new scores are plotted on the updated web in Figure 10-8, which also shows the baseline web for comparison.

If you compare the new web with the baseline web, you will see that some elements have changed. Figure 10-9 summarizes these changes.

Element	Score	Comments
Leadership	12	The new leadership has empowered the organization through the creation of teams.
Work Process	10	The work process has been redesigned, but has not yet been implemented.
Structure	11	Team structure has been created, but the transition is difficult and slow.
Group Learning	5	There has been minimal group or individual learning. The process is just beginning to change.
Technology	10	Adequate. No changes yet, but some is needed in the future to support the new process.
Communication	4	Although the new leadership has made significant strides, it has not effectively communicated this progress to the organization.
Interrelationships	8	Although this area is slowly beginning to get better, there still is a degree of mistrust based on past experience.
Rewards	9	Individual performance rewards, none for teams or groups. This has not yet changed.

Figure 10-7 Reassessment After Three Months

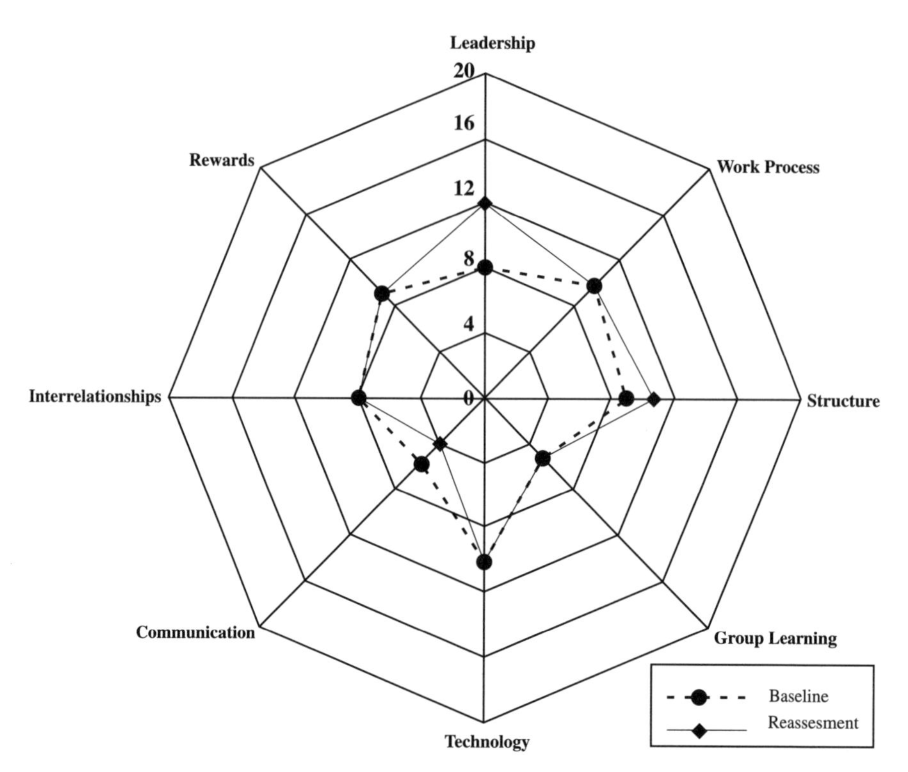

Figure 10.8 The Web of Change with Baseline and Reassessment Scores

Element	Baseline	+ Six Months	Difference
Leadership	8	12	+4
Work Process	10	10	No change
Structure	9	11	+2
Group Learning	5	5	No Change
Technology	10	10	No change
Communication	6	4	-2
Interrelationships	8	8	No change
Rewards	9	9	No change

Figure 10.9 A Comparison Between Baseline and Reassessment Scores

The leadership has changed and, along with it, the structure of the company—both in a positive direction. The remaining elements, except for communications, have not changed. This result is not surprising because not enough time has elapsed for significant change in the other areas to has taken place. If you look closely you will see that something else has happened. The element of communication has slipped from 6 to 4. It has not changed for the better; it has changed for the worse. The web shows not only changes over time, but also helps you identify interactions among elements. The change in communication indicates a possible problem. You need to find out what it is and take corrective action, if necessary. In this simple example—one that is far less complex than a real change process—an explanation for this drop is that without the strict hierarchy, people are initially less comfortable. As a result, the communication level has decreased. The survey and web help to highlight this problem. In turn, you can now perform a more detailed analysis and correct the problem.

Be aware that the scores are not created in a manner that makes them statistically accurate. They are directional, providing a simplified way to measure change. They will show you if elements have improved or deteriorated and by how much in general terms. They also indicate if one or more elements are having a serious negative impact on the others. This information does not solve your problems (a score can never do that). It provides direction so that you will ask questions and conduct the more detailed analysis needed to find the solutions. The actual analysis and problem solving falls to you.

10.7 Next Steps

The purpose of this chapter has been to introduce the Web of Change. You will be able to use this tool as you move forward into Part Three. You have learned not only the concept, but also how to create a web of your own. In addition, you have seen a simple example of how to use the data both as a baseline to measure progress and as an indicator of interaction among the elements.

When you complete the survey in the appendix or on the disk, then use the data to create your web, you will have given yourself and those involved in your change process a valuable tool. Remember that there is no statistical precision to the numbers. They are directional indicators of positive, neutral or negative change. They should be used to help identify questions that must be raised. With answers to these questions, you can understand why changes are happening the way that they are and begin to correct any problems.

Part Three

Execution: How to Make Your Effort a Success

Part One introduced the terminology of change, concerns about making assumptions without substantiating facts, and the concepts of vision and spiral learning. Part Two then introduced the processes and structures that are the foundation of the change process.

Part Three now discusses the eight elements of change. These elements were already introduced in the web diagram described in Chapter 10. They are described in further detail Chapters 11 and 12. After reading these chapters, you will be ready to take the web survey, located in the appendix as well as on a diskette that accompanies the text.

After you have completed the survey, Chapter 13 will explain what to do with the scores, showing you how the elements fit together. Chapter 14 then takes your understanding of the elements and addresses how the actual work of change gets accomplished. Whereas Chapter 7 discussed process teams, Chapter 14 discusses project teams. The last chapter in this section, Chapter 15 is about measurement. I believe that what gets measured gets done. Therefore, what better way is there to conclude the execution section than with a discussion of how to measure the success of the effort.

Many of the chapters in Parts One and Two can be read in a random order. For Part Three, however, I would not recommend it. You need Chapters 11 and 12 along with the survey to help you understand the scoring in Chapter 13, how things get accomplished in Chapter 14, and how measurement applies in Chapter 15.

Chapter 11
The Elements: Part 1

11.1 Introduction

Suppose you are given the task of assembling a team to develop and implement a change initiative. Senior management will probably supply your team with the overall vision, and in many cases, the goals to be achieved. What you haven't been given are the details of how to turn the vision and related goals into a specific plan that will provide the change. When the team gets together, you must determine how to accomplish this task. (I will discuss this process in detail in Chapter 14.)

There is an essential set of eight elements that the group must address before it can successfully tackle this major effort. Failure to address these elements can lead to gaps in the plan for bringing change to the organization. These elements were introduced in Chapter 10; they make up the Web of Change. They are

- leadership,
- work process,
- structure,
- group learning,
- technology,
- communication
- interrelationships, and
- rewards.

Chapter 11

Although each element is unique, as a group they must work interdependently if your effort is to be a success. I look at the elements as a structure with four levels (see Figure 11-1). If all of the pieces are properly placed, the structure will last for a long time. If, however, the structure is missing an element or it has been installed incorrectly, then the structure will soon crumble.

Level			
Sustainers	Inter-relationships	Communication	Rewards
Core	Structure	Work Process	
Enablers	Leadership	Technology	
Culture	Group Learning		

Figure 11-1 Visual Model of the Relationship Between the Elements

At the foundation, (the first level) is the organization's culture, that intangible thing that makes you what you are and defines how you will conduct business. Have you ever worked in a department where coming to work was fun or where the group interacted so well that you enjoyed sharing projects and other initiatives? Or have you ever worked in a department where people did not get along and new ideas were confronted with "we tried that and it never worked"? These kinds of behaviors help define an organization's culture. Many times cultural change happens from outside the organization, whether through a buy out or new management replacing those who are leaving. Cultural change can also happen when the existing leadership is faced with the fact that if things don't change the company will be out of business. In either case, cultural change is a major event. Companies that are able to change their culture are referred to as learning organizations. For this reason, I have placed group learning at the base of the structure.

At the next level are the enablers. These elements, leadership and technology, build on the foundation of learning. They enable the group to change successfully. I have seen many thoughtful change initiatives fail because the leadership either didn't understand them, didn't want change to happen, or were comfortable with the status quo. I have also seen changes take place because the leadership strongly believed in and led the change.

Technology is also an enabler. With cutbacks and smaller staffs, both information and ways of completing jobs in less time are important. These enablers—good leadership and sound technology-based solutions—support other components of the change process.

The next level of the structure, made up of structure and work process, is the core. Structure defines how you are organized to accomplish the work whereas work process defines how the work actually gets accomplished. When you develop your change initiative, these two elements are essentially what you are going to change.

Once you have worked through the core elements, the remaining elements are the sustainers. The long-term success rate of any change is dependent on communication, interrelationships, and rewards. Communication focuses on how information is transmitted from those who have it to those who need it. Interrelationships determine how well people within the company get along and work cooperatively toward a common goal. Good relations can not by themselves overcome other problems and deficiencies. However, bad relations can undermine the best of plans. The reward system is the last of the sustainers. In many cases, the reward systems of the past will not sustain change. We need to create new reward systems based on a work team approach if we wish to sustain the change we have created.

11.2 Leadership

Leadership is essential for a change effort to succeed. Figure 11-1 identifies leadership as one of the enablers. Indeed it is because without leadership, you are set up to fail.

A change management leader is a person who can clearly see and communicate a vision of the future, influence others to embrace the changes necessary to accomplish the vision, maintain organizational focus over an extended time period, and support group learning as the process evolves. It short, "what we conceive and believe, we can achieve." The leader is the person who must pull it all together.

According to this definition a change management leader needs two sets of skills: transitional and transactional. The transitional part of leadership enables the organization to move from the current way things are done to a new way. Transitional leaders focus on the future and work with the organization to achieve the vision. Transactional leadership enables the organization to take the vision and convert it into something real, something that can be accomplished. What a company wants, but does not always have, are leaders with both sets of skills. Figure 11-2 illustrates how these two sets of skills fit together.

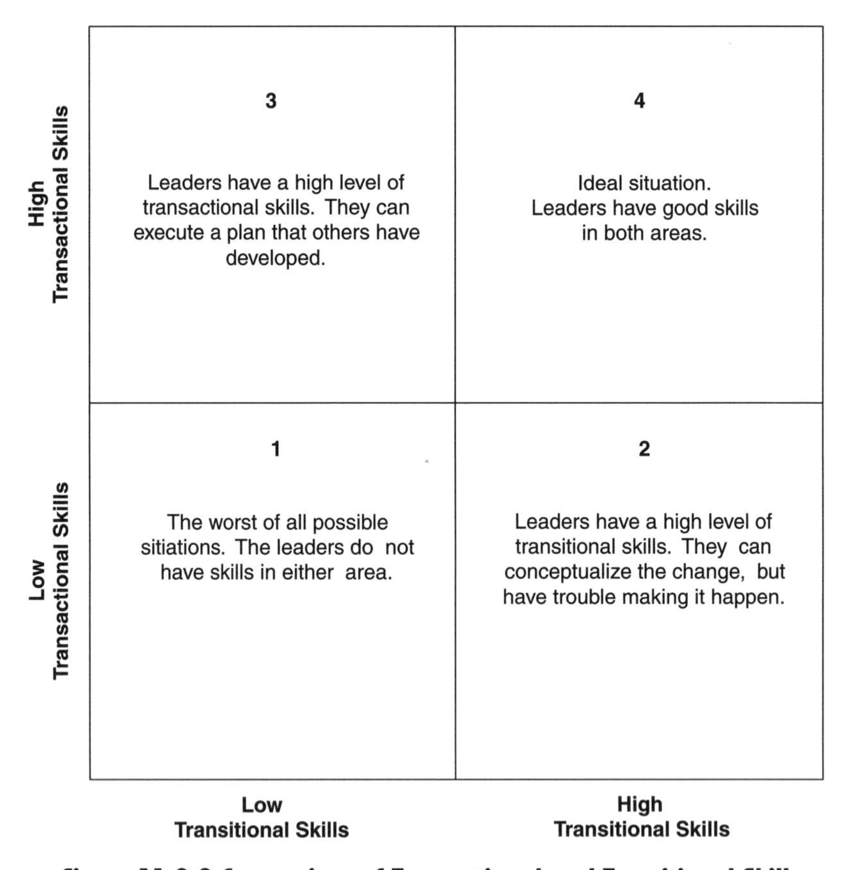

Figure 11-2 A Comparison of Transactional and Transitional Skills

In this figure, the *y*-axis represents the transactional skill and the *x*-axis the transitional skills. Quadrant 1 is unacceptable. If your leaders have low skills in both areas, your change effort is in trouble. The leaders can not take you into the future. They have equal difficulty working with day-to-day transactions. In Quadrant 2, the leaders have high transitional skills, but low transactional skills. They can conceptualize change, but have difficulty taking a concept to reality. This problem can be overcome if they realize their shortcoming and bring in other people who can convert conceptual ideas into actual work. In Quadrant 3 are those who are good at task-related work. Given an idea, they are the ones who can make it happen. Although they are good at transactions, they are not good at visualizing new and different concepts. This problem can also be overcome if they bring in people who can see and articulate new ways to do work. Quadrant 3 leaders must also be willing to accept the plans of others. Then the change process can move forward. The last quadrant is really where you want to be. In Quadrant 4 are the leaders who have both sets of skills. They are complete change management leaders. These leaders, however, are rare. The alternatives in Quadrants 2 and 3 are far more likely. If your leaders fall into the Quadrant 2 or 3 and are willing to be driven by purpose rather than ego, they can form teams with the proper skills to make value-added change happen.

Leadership is not found only in the single person at the top of the organization. Yes, leadership is needed at the top. However, leadership is also needed throughout an organization for real change to occur. The reason is long lasting change takes place over a long time. Companies are not transformed in a week. They are transformed over years of hard work. The effort to bring about change is not confined to one overall leader; it must come from all levels of the organization.

The following expectations describe the characteristics you and your colleagues need have as leaders of your own change processes:

- Inspired, with a strong belief that change is required for success
- Visionary in that you see and understand the vision
- Confident that you have or can obtain the skills to help the organization achieve its goals
- Committed to stay the course and overcome barriers
- Empowered to act and willing to empower others to do the same
- Comfortable with others who have skills that you do not have and smart enough to get them to help
- Willing to say "I'll try" and unwilling to accept "I can't"

11.3 Work Process

Work processes are a core element of the change process. When you make changes, the processes are what you change. Some processes are used throughout a business, some are purely functional, and others cross departmental lines. These processes illustrate to the outside observer how work is supposed to flow through the system. The key phrase is "supposed to." In addition to the official work processes, there are processes at work behind the scenes. They have evolved to handle deficiencies in the system that the organization has not yet addressed. As you develop a plan for change, you must address both the official and the behind-the-scenes processes. You are then more likely to address the deficiencies that need to be resolved.

Flow Charts

Many work processes are shown in the form of flow charts. Some flow diagrams are coupled with detailed explanations of each step or block of steps; they may also list who has the responsibility for the step. If you don't already have your key processes diagrammed, you should take the time to develop these flow charts. Start by making a list of steps, prioritize them based on importance, and then look at the sequence. This effort will help you understand the work process. You may even learn some things about your workflow that you didn't already know.

As you prepare a flow chart, I would like to suggest that you use the simple set of symbols illustrated in Figure 11-3.

The symbols in Figure 11-3 are ones that I use for virtually every diagram that I create. You may choose to use them or select a modified set that suits you better. An example of a flow chart using these symbols is shown in Chapter 16; it describes how to audit the change process and correct identified problems.

Keep the following points in mind as you and your team develop work process flow charts.

- Get as complicated as you need to represent your workflow accurately. If you develop them at too basic a level, the really important details and potential improvement areas will be unrecognized and lost.
- Number each block. Then in textual form you can more fully describe that part of the process. Numbering blocks will help you reference these details as you work your way through the flow chart.
- Include the job title responsible for each task in the accompanying detailed description. You can then keep track of who is responsible for each task. You can also look at the interfaces between job functions and groups. Use job titles instead of people's names so that the flows remain current long after people have changed jobs.

- Use multiple pages as needed. Capture all of the information which many times may not all fit on one page. Use the off page connectors identified in Figure 11-3.
- Incorporate group knowledge as you develop the flow chart. If you prepare the chart in a vacuum, you will miss important details that a group would include.
- Mark the chart with the date it was created and its revision number. Many times I have reviewed multiple versions of the same chart. The revision number and dates prevented a lot of confusion.

Creating flow charts that accurately depict a work process is a skill that needs to be learned through repetition. Work with different groups to incorporate their understanding of the workflows into the chart. Correct it until they agree that the chart accurately reflects the actual work process-

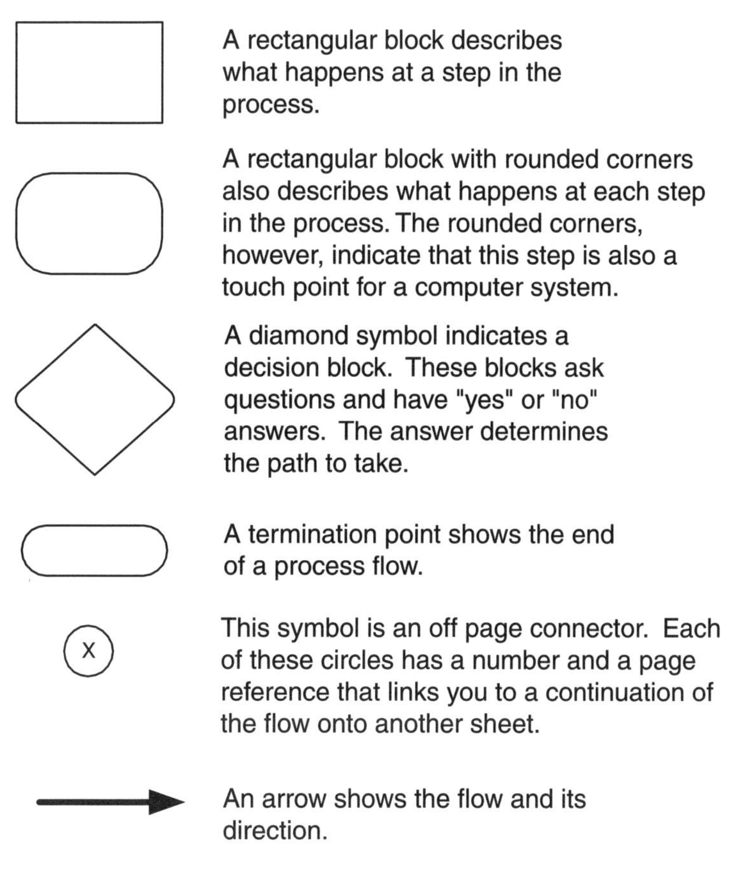

Figure 11-3 Work Process Symbols (A Simplified Version)

es. As you learn to develop these charts, learn which questions get at the information you need. Your charts will be valuable to others, allowing them to see pictorially what they do so that they can critically review their efforts and improve upon them.

Types of Work Process Change

You should now be able to develop the diagram of your workflow. The work process is the main area where you would look when planning improvements for your company. Chapter 14 will discuss more specifically the steps you can take to plan these changes. First you need to understand what you currently have.

Figure 11-4 indicates various types of change that you may consider as you begin planning how to make improvements.

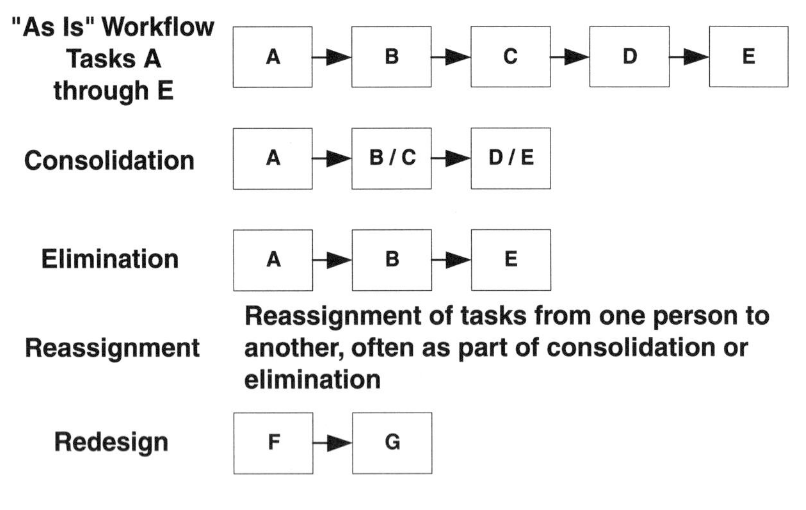

Figure 11-4 Types of Change

As Is. This sequence describes the process as it currently exists at your company.

Consolidation. This change involves combining tasks done by one or more people. It reduces potential problems at the workflow interfaces, where work is passed from one person to another. By involving fewer employees in the process, it creates added ownership for those remaining.

Elimination. With this change, you actually remove steps in the work process, essentially streamlining the effort. If you have created a detailed flow chart, you will more easily see steps that can be eliminated. A word of caution: Be sure that you fully understand what steps you are removing from the process. You don't want to learn later that your new process has eliminated a step that is of critical importance. A good way to prevent such an error is to interview those involved with the actual work. They will be the first to tell you about any risks of eliminating steps from the process.

Reassignment. This change is similar to consolidation. Work is transferred to someone other than the person currently responsible for it. Moving a task to a different department can often improve the process.

Redesign. This type of change does not tinker with an existing process. Instead it creates a new and different way to handle the work. Although redesign can add the most real value to an organization, it is the most difficult to accomplish.

When you affect the work processes, you also affect people. What people do, how they do it, effectiveness and efficiency issues, communications, training, structure and interrelationships all must be fully addressed when you change the work process. Failure to think through and address these issues at the outset is a serious omission, it will result in problems or outright failure of the process itself.

11.4 Structure

Structure is the second of the core elements. As you work through the work process changes, you often arrive at the realization that you need to change the organizational structure. Why? Because structure follows strategy. A redesign of the work process drives structural changes that support the new strategy.

There are many types of organizational structures and many texts on the market that describe them. What works well now for your organization may not be the best once your work processes change and the company evolve. As you consider how and if to restructure, keep in mind the following:

- Only restructure if you have a valid business reason. Restructuring is unsettling to employees. Employees worry, often for good reasons, about losing their jobs during restructuring, even employees who provide high value to their companies. If you must restructure, then communicate the business reasons as soon as possible. This may not totally remove the fear factor, but it will help employees better understand what is going on.

- When you change a structure, clearly define expectations of the new jobs. This will help smooth the transition.
- Think out of the box when you develop a new structure. You want to be able to redesign, not just shuffle the deck.
- Try to ignore who will be doing what. Don't assign names to the blocks and don't assume that you know who will get a specific job. Focus on the design itself and the list of job expectations.
- Make sure that what you design supports the work process. Use the process to help you with the structural changes so that they are mutually supportive.
- Don't change too often. Frequent change destabilizes the organization, leading employees to think that you don't know what you are doing.
- Provide adequate resources to cover vacations and other absences.
- If you are in a union plant, factor in contractual issues.
- Create the new structure so that employees can't return to the old ways. Otherwise, they will seek to reestablish their comfort zone by abandoning the new and recreating the old.
- If you are creating a team-based organization, consider locating all of the team members in the same area—even if they are not in the same department.
- Build a structure that makes it easy for work to move across interfaces with minimal difficulty.
- Don't create a new structure by yourself. Work with a group and include those who will be impacted by the work redesign. This can be difficult because people want to protect their turf.
- Communicate what is going on early and often. Restructuring frightens people. The more you let them know, the easier the change will be.
- Clearly understand and address all training requirements so that employees with new jobs or new responsibilities know what they have to do, and know how to do it.
- Develop a detailed transition plan and follow it.
- Evaluate the change in structure after it has been in place for six months. See if it has accomplished your goals. If the transition looks good, then reevaluate again six months later.

Structural and work process changes are the most significant elements of change. They have the greatest impact, not only on the work, but also on the employees. Changes in these areas are often necessary, but proceed carefully to assure success.

11.5 Group Learning

Typically, when we talk about what is needed to help employees work more efficiently, we immediately think about training. Organizations often ignore group and individual learning. One type of training is designed to help employees improve their skills. Examples include training mechanics how to layout piping better or machinists how to repair a pump more efficiently. Training even applies to managers. Training programs are usually developed by analyzing job requirements, comparing them to current performance levels, identifying performance gaps, and training employees to eliminate or reduce these gaps to an acceptable level.

Training is also important for other reasons. Skills can deteriorate through lack of use. New employees entering the organization may not have all of the skills needed. The nature of the work may change over time. Training keeps employees proficient in the work place as it is.

As important as training is, it does not help an organization that is undergoing change move forward. For such organizations, tasks that were done one way may now proceed in a different manner. A different kind of learning is needed for change, one that goes beyond simply training new skills. Employees must develop the ability to learn as they progress along the spirals described in Chapter 3.

Spiral learning is not as difficult to apply as it is to support. The hard part for you as a manager is accepting that although your employees have answers for the first spiral, and maybe think they know what is going to occur in the second spiral, they may not know what will happen over the horizon. From the manager's perspective, having work proceed using the spiral learning process can be uncomfortable. However, it is worthwhile, leading to great success. Furthermore, forcing linear thinking on a change process can disrupt both the process itself and the organization's ability to learn as the process evolves.

Employees also need to learn what have often been referred to as the soft skills: communication; interrelationships, supervising, assertiveness, and teamwork These skills are often overlooked because they do not provide the obvious payback or return on investment that traditional job training provides. Yet without these skills the spiral process will not succeed.

Individual learning of these skills is not enough. Employees need to learn in groups. Earlier in this book, we have looked at the value of teams and the power of the group. When employees learn together, their power is multiplied. They progress collectively and, consequently, can deliver a better product.

As a leader, your job is not only to put processes in place, but also to see that you get value from the effort. The steps include the following.

Understand the value and benefits that can be achieved. Management must believe that group learning and soft skills can add value. Management support will lead not only to commitment, but also to funds being made available to support programs that will develop these new skills.

Identify what is available and have employees suggest what they need. Having employees list their needs may seem risky; management may fear it will get a large laundry list of courses that it will not be able to satisfy. Nevertheless, it is a good approach. Employees know what they and their team need.

Develop a comprehensive plan. Once you have your list of general areas and specific skills needing improvement, then develop a comprehensive plan. Because you can't teach everything at once, the plan will be long-term. Developing skills that ultimately improve group learning requires time, not only for developing and conducting the programs themselves, but also allowing employees time to absorb what they have learned. Furthermore, many areas of learning require other learning as a prerequisite. For example, before you can perform team skills, you must first develop the communication skills of the team members.

Ensure that the learning is taking place. Once you begin to execute the plan, ensure that learning is taking place. Make sure that employees understand the importance. Otherwise, they may not even show up or pay attention while the training is being conducted. In addition, give employees time to learn. If their job responsibilities are such that they can not get away, then they may ignore the learning sessions. Companies often address this problem by conducting the training off site, not to provide a nice hotel environment, but to get employees away from their regular jobs so that they can focus. By being off site, the company is forced to cover the absences as if the team members were on vacation.

Ensure that the learning is being utilized. Once the learning programs are completed, you need a process in place to ensure that the new skills are sustained. Otherwise, the perception will be that the new skills aren't really very important; the status quo will again take over. Develop initiatives that apply the new learning. Then, as manager, talk to employees about the importance of the new skills. Alter your own behavior to demonstrate that you too are applying the new skills as well.

Use a facilitator or coordinator to keep the effort on track over time. Moving an organization to a higher level of learning is a full-time task, not a parttime one. It requires a great amount of coordination. Using a coordinator doesn't reduce your ownership. It does mean, however, that you will have a second person helping to make sure that the effort begins well and continues successfully.

Chapter 12
The Elements: Part 2

12.1 The Next Four Elements

In Chapter 11 we looked at four of the eight elements of change. This chapter looks at the remaining four elements: technology, communication, interrelationships and rewards. Following this discussion, I will introduce you to the Web of Change survey. The detailed description of how to complete the survey is in the appendix. I recommend that you take the survey at the end of this chapter. Your information about the eight elements will then be clear. Having the survey results will in turn, help you with Chapter 13.

12.2 Technology

The word *technology* has broad meanings. For this discussion, however, I plan to use it in a narrow context limited to change efforts and how those efforts are supported. Thus technology refers to the software tools that enable and support your business's work processes.

Why is technology so important that it is considered a separate element on the web? It streamlines the work process, captures and retrieves vast amounts of data easily, supports common practices, and enables wide dissemination of information.

If you have been in the work force for more than 15 years, you know that most of what you once did manually is now done faster, more accurately, and more easily than before. For example, consider computerized maintenance management systems (CMMS). These systems, maintain historical databases, retrieve data, process work requests, route these requests to predefined approvers, handle material requirements, pay the workforce, and handle many other functions as well. In years, such work was labor intensive, less accurate, and slower to complete. Technology changed these tasks.

I have been involved in many systems initiatives and there are things that you need to know if you are going to get use technology to support your business change process. If done right it most certainly will help you and if done wrong it most certainly has the potential to be you worst nightmare.

Is Technology a Driver or an Enabler of Change?

There are two different schools of thought about the role of technology in the change process. Many in the systems community believe that the system should come first and the work process should follow. This approach has a significant problem. Suppose you identify the process you want to change, then find a system to handle it. Systems are generally sold by sales representatives who are primarily motivated by selling software. You could get a system that looks good on the surface. However, when you implemented the system, you find it has deficiencies that hinder the change process. Thus, if you start with systems, then any deficiencies in the system become deficiencies in the process. For example, you install a materials management system that is weak in the area of procurement. When you install the system, you will be actually implementing this weakness into your work process. In turn, the users will be extremely dissatisfied with the results. They may, with just cause, resist the implementation.

The alternate approach is to develop the work process first, then find software that allows you to make the process work as you designed it. At first, this appears to be the better way to approach change. The work process is developed to your vision so that it has no deficiencies: you have created what you want. Next you need to find a software that supports the new process. This approach not only removes the deficiencies, but also allows you to better judge what you are buying. Rather than getting canned demonstrations from the vendor, you are getting the software vendor to demonstrate how the system fits the various work processes you have developed. If functionality is missing, you can continue your search. If you decide to select the software, you can work with the vendor to get the functionality improvements added to a future release. You can also develop a way to use the existing functionality to accomplish what you want or develop an alternative.

This approach has its own pitfalls. First, you need to develop specific selection criteria for the software vendor to use in a demonstration. The vendor needs to be able to demonstrate all of the various functions that you need. In some cases, mostly including smaller systems, you can even have the vendor let you work directly with the system. Remember that the vendor's sales representatives are rewarded on sales; they may tell you that something can be done when, in reality, that functionality does not yet exist. Be careful with promises about what will come in the next release, commonly referred to as vaporware. All too often, software that exists as vapor remains as vapor.

The other major pitfall if you are developing the process before finding the system, involves your employees. You will have many committed employees waiting to see how well software supports the processes that they developed. If it does not deliver according to their expectations, you will have many unhappy workers. You can manage expectations, however, by involving them in the selection process. Then when the final selection is made, any deficiencies are already known. Adjustments to the process can be developed more easily and with more support. This approach also ensures that technology—in this case the system software—is enabling the change effort, not driving it. The change effort is still driven by a thoughtful analysis of the company's needs and its processes, not by the convincing arguments of a sales representative with a product to sell.

Generating Reports

The implementation of systems software is important. Of equal importance is the information that you get out of it. Systems today store huge amounts of data. The challenge for users is retrieving the data and transforming it into useful information. Most software systems provide "canned reports" in a predetermined format that you can extract as needed. However, the reports that are available may not provide what you want. The format may be different and the data or information itself may be different. This discrepancy causes problems, frustrating those who are trying to use the system

The issue of what reports your system software should generate is a difficult one. It must be resolved by the project team that will implement your system. Do not, under any circumstances, allow the team to avoid addressing this issue in great detail. Any system must have the reporting functionality that you need to conduct your business. Make sure that you see it, feel it, test it, and are fully satisfied with it. When you analyze the reporting functionality, keep the following points in mind:

- Understand exactly what reports the users need. You will then be able to ensure that the system you select can deliver them. Ignoring user needs and assuming that the system will deliver the necessary reports is a certain way to undermine your efforts. I can assure you of this from personal experience.
- Know completely what the system can deliver. Standard, formatted reports may be useful at times. However, customized, user-created reports will be needed too. Be sure that the system has the flexibility and capability to deliver the reports in the way you conduct business.
- Learn how difficult it is for the average users, not the computer experts, to get the information they need whether they just want to create a simple report or want very detailed complex reports.

What Is the Return on Investment (ROI)?

When you request funding for a technology (or any) project, management will want to know what return on investment (ROI) can be expected for the money that will be invested. Suppose you want to install a new piece of equipment that costs $100,000. This equipment will generate additional revenue at the rate of $20,000 per month. The simple return on the investment (not factoring in the time value of money) is $100,000/$20,000 or five months.

Calculating the ROI for systems software is a bit trickier. Systems seldom drive revenue the way a new piece of equipment does. Systems are enablers. Therefore when you install a system, you enable something else to happen that should generate revenue for the company. If, however, the process isn't managed well, then the system is nothing more than a very expensive piece of software. How then do you respond when management asks for the ROI of a $500,000 investment for new software?

First be sure that management understands that simply installing software will not generate revenue. Management needs to understand that if it gives you the money for the software and does nothing else to support a change effort, the money will have been effectively thrown away. Second, provide a detailed plan that shows how using the enabling system software will drive benefits in other areas. If you do not show this connection, you will probably fail to get the funding you seek. You may be competing for funds against other projects that may directly drive benefits for the corporation.

Suppose you want to buy a sophisticated piece of software that costs $500,000. For this investment, you will get functionality that will support an important process that is being redesigned. When you ask for funding, management asks for the ROI. You can not respond that the software is just an enabler of change and that it doesn't have an ROI. Nor would that be accurate. Instead, you need to explain in detail both the cost savings and revenue enhancements that will result from the software being purchased and installed. For instance, the system may allow you to eliminate a staff position through attrition. The salary and benefits you would have paid that employee is a cost saving that factors into the ROI. The system may help you better monitor raw material inventories, decreasing downtime by ten percent—another component of ROI. It may also streamline the sales force's reporting requirements by ten percent giving them more time to sell to new clients, with still more impact on the ROI. In short, make the connection of how investment in the software will enhance your company's profitability.

If you succeeded in getting the funding you requested, deliver what you promised. This applies not just to the technology itself, but also to the enabled processes. Even if you haven't been given direct responsibility and accountability for these enhanced processes, others have been. They need to make results happen. Unfortunately, many systems authorized without a detailed follow-up plan never deliver the nontechnology benefits. Having a detailed plan is your best guarantee of success.

Acquiring New Technology

You should follow a clear set of steps when identifying the system software you need and the vendor who will provide it. With slight modification, the steps described in Chapter 8 for selecting a consultant apply to systems software as well.

Chapter 12

Step #1 Recognizing the Need for Software

You know you need new software when the existing software can no longer provide the functionality you need to conduct your business. Understanding that your software is outdated may be hard to prove to those who don't use it often. These same people are often the ones who you need to authorize funding. One factor that often forces a change occurs when the vendor who supplies your software no longer supports it. This may seem trivial, but it isn't. With an unsupported system, if you run across bugs or need support, the vendor will not easily be able to provide it. Whether or not this is a reason, you need to be able to show how a new version will enable additional value for the company.

Step #2 Detailed Scope

Before you proceed further, you must develop a detailed scope. Otherwise, you really haven't taken the proper time to figure out what you need. With their broad-based knowledge, consultants can help you in this effort. Even if you don't use consultants, you need the scope to proceed. It will help you to identify and get what you need.

Step #3 Finding the Software that Can Help You

Finding new software often assumes that you are a first-time shopper. Those who have bought and implemented software before seldom look for a new vendor unless there is a problem with the existing vendor. It is just too expensive. Instead they tend to upgrade to new releases of software that is already in place.

For new software, trade shows, trade magazines, consultants with systems knowledge, and referrals provide the best route when you need to find out what is available. In some ways, finding new software is less difficult than finding a consultant; the number of available software applications that fit your requirements are fewer. However, if you make the wrong choice, you can generally replace consultants more easily than systems software.

Step #4 The Presentation

Once you have decided on potential vendors, you need to bring them in for presentations of their products. Make sure that you have told them the functionality you want to see. Then make sure that you see it. Avoid vaporware.

Step #5 The Short List

After the presentations are over, create your short list. This is the list of the vendors you might select, eliminating those you will no longer consider. At this time, there are two actions you can take to help your decision. First, visit a site where the software is already being used. The vendor should be happy to take you to a client who is already using the software so that you can observe how well it will suit your needs. Second, visit the vendor's home office. This trip will allow you to meet some of their employees and get a better sense of the firm itself. This research may require travel. However, it is necessary to help you make the correct choice.

At this time, you should also consider cost. Most vendors have specific ways that they cost out their products. You need to understand each one. Because the costing methods can be different, develop a review criteria that compares them in a similar fashion.

Step #6 The Second Presentation

Now that you have done all of your homework, have the vendors make a second presentation before you make your final selection. This one should be more detailed than the first, allowing you to gain a fuller understanding of the product. This also allows you to get any additional questions answered.

Step #7 Final Selection

Once the second presentations are complete, you and your team can decide which vendor you will use. Remember that you can not simply fire systems software if it doesn't perform. The selection of software is like a long-term marriage. You want a good working relationship with both the vendor and the software. Some great vendors have software that doesn't match your needs. Some great software is supported by vendors who won't properly support you. The right combination is important to the business. It is also very expensive to change if the software or the vendor doesn't live up to expectations.

Chapter 12

Implementing Software

Once you have selected the software your work has just begun. Implementing new software is an area where the right consultant can be of immense value. They have gone through the process before and know many of the pitfalls to avoid. Because this type of project can be very expensive, you don't want to learn by trial and error. Take advantage of what a consultant can offer in areas such as hardware, linking the software to the work process, training, and reporting.

If you are starting to build an integrated database from scratch, you will find that the data you want is all over the place. Some may be in an old software system, in spreadsheet files, or even in hard copy in people's desk drawers and file cabinets. This information must be converted into a new integrated database. In addition, if you are installing a system that has an equipment or materials database, you should go out to the field and validate what is there to make sure your data is up to date and accurate. Such work is time consuming and expensive. You may think that your own personnel can complete the job. Although they probably can, go with a specialist who knows what to look for and how to efficiently go about the task of data gathering.

12.3 Communication

Communication is one element that sustains the change initiative. Without good, clear communication, you will find it difficult to get anything accomplished. But communication isn't as simple as black and white. There are all shades of communication, both good and bad, that support or hamper change efforts.

Several years ago I was given my first supervisory assignment. I supervised three foremen, a planner, and about forty mechanics in a maintenance area within the plant. When I arrived, the shop area where the group worked was very disorganized. I had some ideas about how to change it. However, I wanted to proceed slowly, getting a sense for the group before I made any major alterations to the way they were working. I mentioned to the staff my desire to improve how things were set up in the shop. On the surface, this seemed like a fairly simple communication. It was not.

My intended message was to alert them that at some time in the future, we were going to change how the area was organized. What I didn't take into account was my staff's frame of reference. Their former supervisor was a virtual tyrant who expected that the simplest suggestion be carried out instantly. After my seemingly simple comment to the group, I went to a meeting. When I returned several hours later, the entire shop area had been taken apart. The work group was cleaning and reorganizing the area. The group's daily work plan for the entire day had been disrupted. My simple communication was misunderstood, resulting in a highly disruptive and unplanned event. Needless to say, I didn't start my new job under the best of terms with those who had expected the work group to make repairs on their equipment. The problem occurred because what I said is not what was heard. Making sure that the message sent is the same as the message received is central to effective communication.

The Communication Diagram

Figure 12-1 illustrates the many facets of communication. They must all work together to ensure good communication, whether you are sending or receiving information. Essentially any communication has six

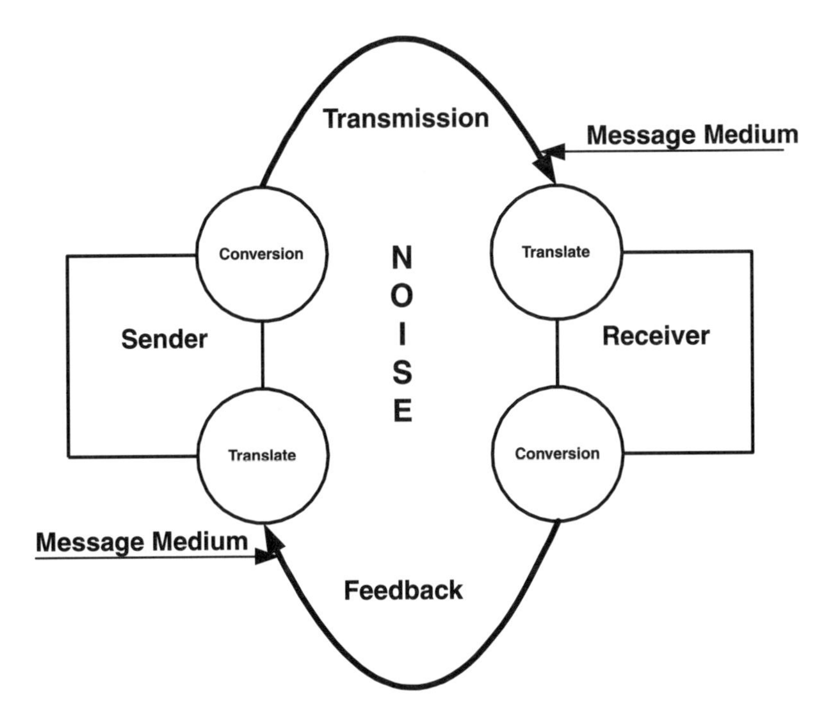

Figure 12-1 The Communication Model

components: developing the thought that you want to communicate; converting that thought into a message; sending the message through various media; translating the message by the receiver; receiving the message; and feedback. Each has its own set of pitfalls, places where communication can get fouled up and the wrong result emerge.

The Sender. The sender is any person or group who needs to communicate with someone else. The communication could be complex strategic information or an assignment of a simple task. You send information a thousand times a day. This is certainly true in business, especially when you are trying to make a change within the organization. Thus, communication begins when, as the sender, you have a thought to communicate.

Conversion. You must now take your thought and convert it into a message that you can send to the person or group you want to receive it. Converting a thought into a clear message is often difficult. Think about any time you had to organize your thoughts for a presentation to a group. The process was probably difficult. Even after it seemed you had completed the process successfully, you found that some people in the audience still didn't understand what you meant. The key to successfully converting thoughts into a message is to take the necessary time to make the message as clear as possible. The better that you prepare your message, the more likely is that it will be properly received and nothing will be lost in the translation.

Transmission. Once your thought has been converted into a message, you need to send it. This may sound easy, but it isn't. There are many ways to send the message. How you choose to send it can be as important as the message itself. Media run the whole gamut including oral transmission—speaking your message face to face in a one-on-one setting, speaking in a group setting, talking over a telephone—to written transmission—personal letters, group memorandum, and computer e-mail. Transmission can be in real-time, such as a telephone or instant messages—or delayed time, such as voice mail or letters. If you select the correct medium, you have a higher chance that your communication will be successful.

Suppose that you are ready to roll out a new and somewhat complicated work process change in your firm. Would you send everyone an e-mail? Probably not. You would not choose this medium because not everyone has the same ability to receive it. Furthermore, the levels of understanding would be varied, and in some cases not what you wanted. Instead, you would bring small groups in for training, using the medium of small group discussion to communicate your message. In this way, you have the ability to interact at a personal, face-to-face level, seeing directly that the employees actually understand what you mean.

Translation. When the message reaches the receiver, potential problems can still exist. The message at this point needs to be translated into something that the receiver can and will understand. Unfortunately, the message isn't always translated into the actual information that you wanted to convey. The receivers frame of reference can cause the mis-translation. Certain phrases may mean one thing to the sender, something else to the receiver. Spoken words can be emphasized in a different way than written words. Sometimes, translation is really the problem: for example, attachments to e-mails often turn into gobbledy gook by the time the recipient opens the file. The feedback stage provides an important way of determining that the translation was clear and accurate.

Receipt. The last part of the trip for the initial communication is its final receipt. This is the step when the receiver processes the message into action. If everything has gone according to plan, then the receiver has received what was sent and acts accordingly. However, if the process is anything like my first supervisory experience, the message is incorrectly received and inappropriate action is taken.

Feedback. The receiver has a responsibility to assure that what was sent was received. This check is done through the process of feedback. Simply restating the message to the sender usually provides this check. The sender can then either confirm that the message was received or provide the receiver with clarification. If you think about my supervisory example, the staff could have asked me if I meant that they should reorganize and clean up the shop immediately. I then would have recognized that my message was not properly received and could have corrected their understanding.

Noise. The last component of the communication process is not a direct part of the sequence. Instead, it exists around the process and is referred to as noise. Noise refers to everything that is external to the actual communication process, but interrupts or disrupts it. Noise can be simply sound, whether static during a phone call or people talking during a presentation. It can also include office gossip that disrupts the message that people receive or routine interruptions as you do your work. All of these make communication more difficult. If you have ever tried to have a serious discussion about a complex work initiative while also involved in the day-to-day hustle and bustle, you may have found communication difficult if not impossible to carry out. However, in an offsite meeting with the key people in the room and the doors closed, you were able to plan the initiative with little difficulty. What was the difference?

Communication Goal

Understand the potential difficulty that surrounds the communication process. Make sure that the messages that you send are fully understood. This applies especially to the area of change management. The process of change is already challenging. When you add the problems associated with successful communication, the process can become even more difficult, with an increased risk of failure. As you begin change initiatives, remember to pay careful attention to what you communicate, how you communicate it, and how it is received. Say what you mean, mean what you say, and then make certain that others understand clearly what you intended to communicate.

12.4 Interrelationships

Every organization has a formal structure that typically dictates the relationships among organizations, departments and employees. Yet that is only part of the story. Beyond the formal structure that we see on the organization chart lies a very powerful set of complex interrelationships. At times, these can be even more powerful than the formal ones. When these informal relationships are functioning well, it is amazing what can be accomplished. When they are not, the opposite is equally true.

Many years ago, I worked for a manager who had a very simple formula for creating and then maintaining positive interrelationships. His premise was simple. He believed that if you treat people with dignity and respect you would receive the same in return. As a result, you could build sound relationships. When I first started working for him, I thought that this perspective was management rhetoric, just another set of buzz words.

This was not the case. In fact, he lived his life and managed our department based on this premise. I watched as he built good relationships throughout the organization where poor ones had existed before his arrival. I also watched as he used his philosophy to build a strong management team and equally strong relations with our union. Although I had been skeptical, after working for him for three years, I embraced his concept. It works. I have lapses when I drift away from this basic philosophy, but time has proven that it is a sound approach to building strong interrelationships among groups and people.

Interrelationships can be vertical or horizontal. Each is equally important for successful change efforts. Vertical relations are those between managers (senior, middle, and front line) and between managers and the workforce. Many of these relations are dictated by the company's organization chart, but there are also strong unofficial forces at work behind the scenes. Horizontal relationships exist among people at the same level, but in different parts of the organization. Interrelationships are influenced by the behavior of those involved; the relationship can be either positive or negative. The forces are the same for both vertical and horizontal relations. How these forces are applied is different. The forces include the following.

Communication. How are communications sent and received? Poor communication at either end can be damaging to relationships. It is important to communicate clearly information that should be shared and avoid communicating things that should be private. This is critical in a changing environment. Otherwise, the organization will fill up with misinformation that is usually negative, damaging the relationships you are trying to build. Communication needs to be open and honest. Think about someone who has communicated with you in this fashion and how that communication helped build a positive relationship. Think also about someone who didn't operate in that manner— someone who lied, gave you wrong information, or used you to get information without providing any in return. Was your relationship with that person positive? Probably not.

People's Attitude about Work and Accepting Responsibility. Sometimes when you ask someone to do something, you receive the answer "that's not my job." At other times, the response is "I'll be glad to help you do that." How did you react to these responses? How did you feel the next time you asked the same question? People's attitudes about work and accepting responsibility can build or break down relationships. While attitude may not be one of the elements on the Web of Change. Nevertheless, it is an important part of successful relationships and successful change efforts.

People's Attitudes about Each Other. As we look at how we relate to each other, we should remember that our position on an organization chart, our educational background, and our financial position in life do not make us better than someone else. One person's sense of "I'm better than you are" has the ability to destroy relations. Once ruined, they are at best difficult to rebuild. The easiest rule for maintaining good relationships is the one I learned from my former boss: treat people with dignity and respect.

Realize That the Formal Structure Often Inhibits Work. As managers we need a formal structure to help us accomplish tasks in an organized manner. However, limiting employees to formal routes through the structure can be detrimental. The team approach—bringing employees from multiple organizations together—does not follow the organization's formal structure, but it works. This is especially true during periods of change when merging employee's collective skills is very important. Do not allow formal horizontal or vertical structures to get in the way of team formation and the benefits it provides.

Good interrelationships are essential to successful change efforts. If they already exist they need to be maintained. Otherwise they can deteriorate and so will the likelihood of success of the change effort. If they do not exist, they can be built, though not easily. Nevertheless the effort is more than worth the try.

12.5 Rewards

Conventional rewards provide the way for a company to recognize your performance. For middle managers, conventional rewards tend to be given once a year in the form of a merit review and subsequent salary increase. If we have performed extremely well, we may also be promoted. This section will not focus on such reward systems. Instead, it addresses rewards that support change efforts.

In this context, there are two types of reward structures to consider. The first provides rewards that reinforce the change initiative as it is being implemented. The second provides rewards for long-term commitments to sustaining the change. Unlike conventional rewards, these rewards are linked to the performance of teams, rather than individuals. This makes sense because change efforts are tied closely to team performance.

Rewarding employees with the purpose of reinforcing behavior—in this case, implementing or sustaining change—is not an easy process. As you develop short-term and long-term strategies, keep the following considerations in mind:

1. Rewards reinforce people's efforts. Decide what you want to rein force. Then develop ways to do it. Conversely, do not reinforce

efforts or behaviors that should not be continued.
2. Frequent small rewards reinforce small steps. In turn, small steps make the larger project successful.
3. Link rewards closely in time with the action that they are sup posed to reinforce. Providing a reward months after an event holds little meaning.
4. Rewards related to change efforts usually can not be applied to an individual. They should reinforce team or group efforts.
5. Developing consistent reward criteria is very important. The value of rewards can be lost if employees think that the selection criteria is arbitrary. Inconsistent criteria can cause you to miss those who deserve to be rewarded or to include those who should not be rewarded. In both cases, you will create resentment and undermine your effort.
6. Money isn't always the reward people seek. Sometimes a small token of appreciation or recognition is of greater value. The reward you give should achieve your purpose.
7. Reward in public; criticize in private.
8. Keep the process simple. Bureaucracy will ruin even the most well intentioned effort.

Short-Term Rewards

In the world of change management, rewards need to drive towards a purpose. The yearly merit or profit-sharing reward, has its place in guiding, developing, and reinforcing long-term behavior. However, it will not help you provide positive reinforcement to a change effort. In the short term, you need rewards that are timely, reinforcing small steps towards the change that you are trying to implement.

Suppose you were trying to put a new work process in place to improve the reliability of your equipment. After all development and training was completed, the process was put in place. Then you started to monitor the results. If you rewarded good performance only during the yearly performance review, what immediate reinforcement did you provide to those executing the new process? The answer is small to none. The reward was far removed from the actual events.

For short term rewards you need to think about a short time frame. The rewards should be immediate and visible. Immediacy tells employees that they are indeed on track and helping the new process to succeed. Visibility uses the success of one group as a model for the rest of the organization. People like to be rewarded for what they do. Immediate and visible rewards for successful employees can help motivate those who are not working as efficiently, giving them models to emulate.

Short-term rewards don't have to be financial. They can be simple and inexpensive. They can include simple verbal recognition, a letter of commendation for the file, a plaque, or being named "Employee of the Month." They can include small gifts, a complimentary meal, or a bonus vacation. Use your imagination and arrive at a set of rewards that fit your company. Remember always to focus on what you are trying to achieve—a step in the right direction of the change.

Long-Term Rewards

Rewards over the long term have a totally different focus. Whereas short-term rewards reinforce incremental steps, long-term rewards reinforce performance on a larger scale. The type that you are probably most familiar with is the yearly merit or performance review. Long-term rewards are useful for evaluating an employee's overall development and to ensure that major goals are being achieved. However, there are several challenges to this form of reward.

Long-term rewards don't get completed. Yearly merit programs, as they are often called, have two parts—the review and the pay raise. Management often provides a pay raise (often based on a divisional pool), but the review is not completed. Employees become frustrated because their boss has not taken the time to sit down with them and review their past year's work. In turn, opportunities reinforcement and correction are lost.

Long-term reviews are completed poorly. A poor review process can be as bad as not having a review at all. At some point in your career, you have probably received a poorly prepared review. The reviewer's lack of quality is immediately obvious and demeans the value of your contributions over the course of the year. Again, whatever opportunity you had to reinforce the change effort is lost.

Many managers like to provide both the review and the raise at the same time because it is easier, especially in a large organization. However if you tell employees about their raise first, they often don't hear what you have to say in your attempt to reinforce the change initiatives. If you try to review the change effort and the individual performance first, they don't hear you because they are waiting to find out about their yearly salary increase. To be most effective, the two parts should be separate. The company usually establishes the timing for the yearly salary increase. Adjusting its timing is probably difficult, if not impossible. Instead, reset the time when you discuss work performance.

Performance is only discussed once per year. Employees like to know if they are following the plan and contributing as expected. Reviewing performance once a year simply is insufficient. This is especially true if the employee is not performing in the manner you want. There is no greater failure on the part of management than to have employees sit down for a yearly review and be told their performance is unsatisfactory. They should know well in advance so that they can take corrective action.

One model that I would like to share with you is simple and already used in many companies. You begin by sitting down with each employee and developing at the beginning of the review year as set of mutually agreed upon expectations—yearly goals. These goals become the performance contract. They are also your vehicle for making sure that change initiatives are in place and employees know what is expected. Your next step is to ask employees on a quarterly basis to comment in writing about how their performance matches expectations. These comments do not just get filed in your drawer. Instead, you sit with each employee, review the comments, offer your reaction and either reinforce or correct the performance. Undertaking this review every three months is not easy. However, the rewards to you and the overall change effort far outweigh the time you take. Everything that you accomplish is built on the accomplishments of those who work for you. Keeping them on track and reinforcing their efforts to implement change is important to your success.

Disincentives and Negative Rewards

Disincentives and negative rewards should be identified and, to the best of your ability, eliminated. They work in exactly the opposite way that rewards work. They can be extremely damaging to change initiatives and, for that matter, to anything else you are trying to accomplish.

A disincentive or negative reward is a behavior that is counterproductive to the change initiative you are trying to put into place. Most of them are practices ingrained in your company's work processes. They may be unique to your business, company, or even department. Many are subtle. As a result, they have a way of going unnoticed.

One example is overtime. Many times, overtime is required and is an integral part of the business. However, suppose your overtime has been very high for years. As a result, many of your employees have earned a lot more than their base salary over a long period of time. Many of them have begun living a life style based on the overtime they earn. What happens then if you introduce through a work process change that focuses the plant on improved reliability? Equipment must now be maintained in a state that reduces failure. One outcome of improved reliability is a great reduction in overtime. In this scenario, the workforce loses by helping you make this change. Their reward is loss of wages and their previous standard of living. Negative rewards such as these must be identified and addressed if change efforts are to be successful.

Chapter 13
Fitting the Elements Together

13.1 An Overview

In Chapter 10, you learned how to build a Web of Change. I used a fairly simple example to show how the elements relate with one another.

Chapters 11 and 12, described in more detail the eight elements of change—the actual spokes of the web diagram. Figure 13-1 summarizes the eight elements. Each element is an important component of a change initiative. It is important that you understand what each can contribute. Even more important than how each element acts independently is how the group work together and influence each other. Think about any change initiative in which you have participated. Then think about not only what you changed, but also what other effects took place. When you categorize the full scope of the initiative, you will probably find that all eight elements were addressed.

At the end of Chapter 12, I suggested you take the survey of the elements. (See Appendix A or the disk included with the text.) If you completed the survey, then you now have your own data and are ready for this chapter, which explains the real value of the web. This discussion has two major parts.

Element	Description
Leadership	Direction and development of a culture that supports a change effort.
Work Process	The way in which work gets done.
Structure	The framework for operation, responsible for how employees are organized within the company.
Group Learning	The manner in which an organization and its employees master new ideas and use them, via spiral learning, to improve the work effort.
Technology	Software tools used to gather, analyze, and transmit information to promote data driven decisions.
Communication	The flow of information (timely, accurate, and complete) to the employees who need it in order to be effective.
Inter relationships	How employees interact both formally and informally to complete their work.
Rewards	Positive reinforcement for achieving goals as well as negative reinforcement used to identify and correct mistakes and poor habits.

Figure 13-1 The Eight Elements of Change

First, we will look at an example of baseline data, based on results from the initial survey of a fictitious company. The baseline web has only one set of data points. This example will also be used to introduce you to a set of charts that the web program develops.

The second part of the chapter continues the example, looking not only at the baseline data, but also at data acquired after a change process has been implemented for a period of time. Here we are working with two sets of data points generated over time. As you compare your baseline with the second set of points, you will see that circumstances have changed. The example discusses the meaning behind these changes, as well as what you need to do with the results.

13.2 Charts That Analyze the Survey

Figures 13-2 through 13-5 illustrate four charts that will help you summarize and analyze information from the survey. The enclosed disk will help you generate these charts. Otherwise, you can create them on your own.

Chart 1. In Figure 13-2, Chart 1 shows the Web of Change with the baseline information. The score for each element is represented on a separate axis on the radar diagram. As you move outward from the center, you see the web has five sections, increasing in score value by four points for each section. The outermost section represents the maximum 20 points that you can score per element in the survey. As you move outward from the center for any element, you indicate a higher level of performance for that specific category. Using the web, you can easily to see where you are doing well and where improvement is still needed.

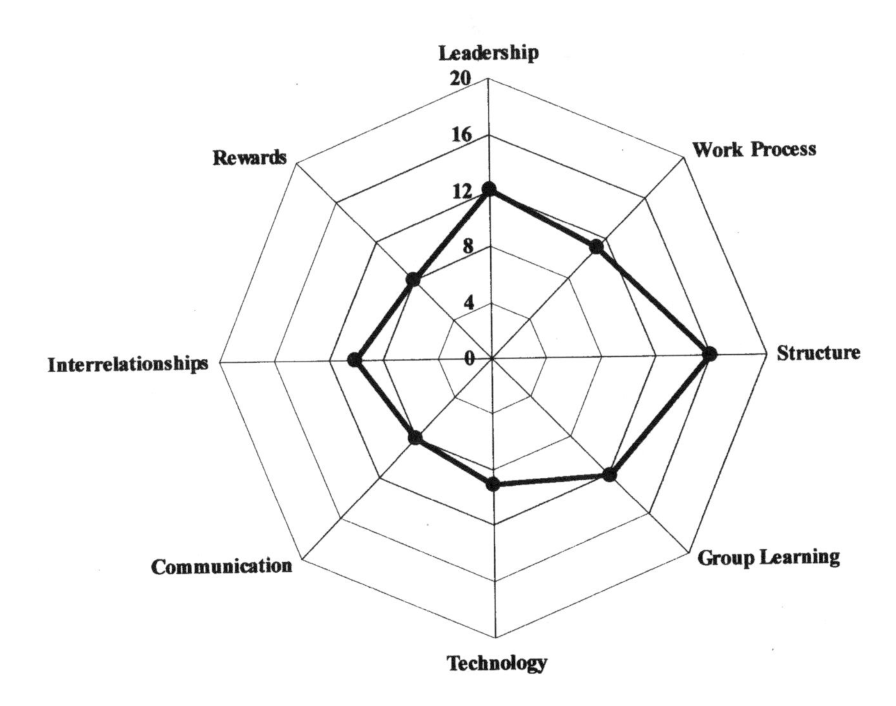

Figure 13-2 Chart 1: The Web of Change with Baseline Data

Chart 2. In Figure 13-3, Chart 2 shows a simple bar chart with the same baseline information illustrated on the web in Chart 1. Chart 1 is designed to show the element scores in relation to each other. Chart 2 focuses instead on the difference between each score and the maximum possible score of 20 points.

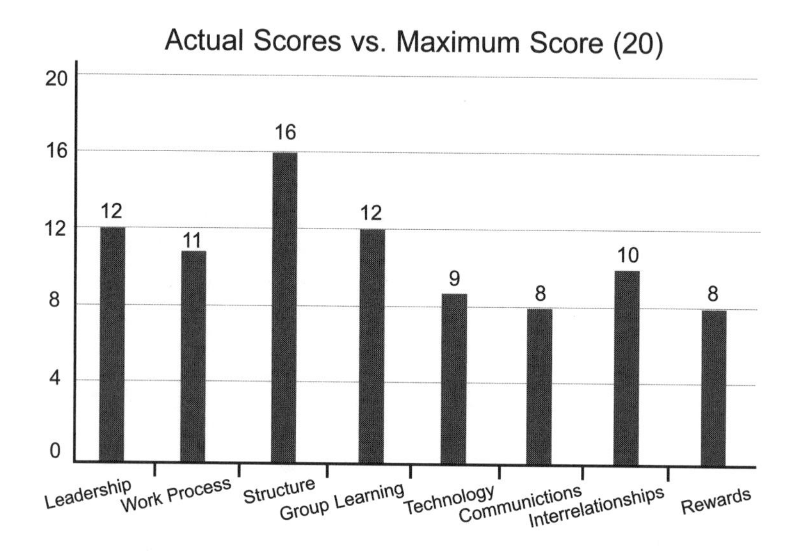

Figure 13-3 Chart 2

Chart 3. In Figure 13-4, Chart 3 compares each individual score with the average of the full set of scores. You want your change initiative to be well balanced, with progress being made over all areas at generally the same rate. Therefore, the individual scores should stay close to the average of the survey. Large deviations, either positive or negative, need to be analyzed. They suggest that one or more of the elements is proceeding

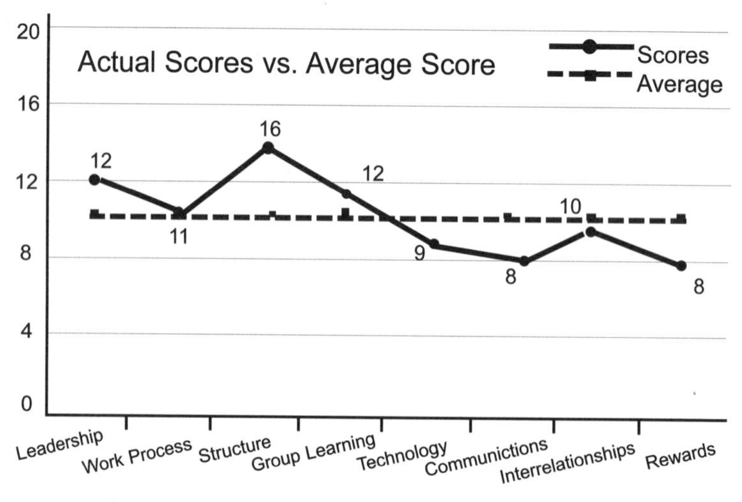

Figure 13-4 Chart 3

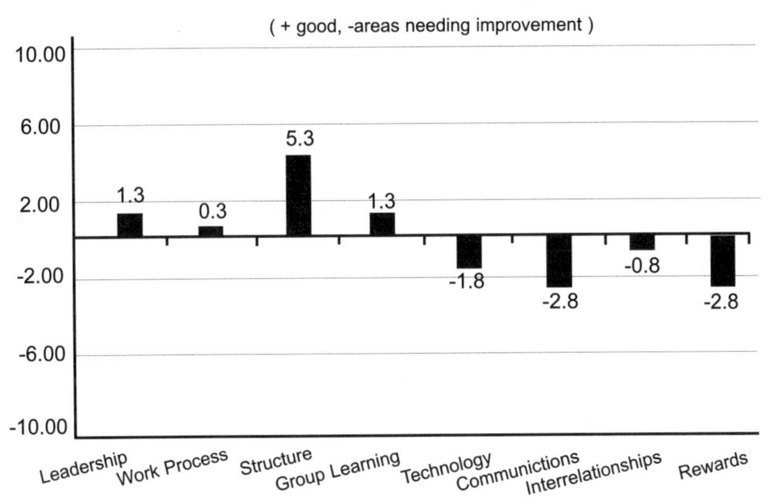

Figure 13-5 (continued) Chart 4

faster or slower than the others. Although the chart itself will not tell you why, it can point you in the right direction.

Chart 4. Chart 4 in Figure 13-5 also shows the deviation of actual scores from the average, but in a different way than Chart 3. Here you are able to look at the differences between the actual scores and the average score differently and more dramatically. In this chart, look for the deviations, both positive and negative, from the average score represented by zero.

Charts 1, 2, and 3 are the same when the data points from the reassessment are added. The difference is that you now can compare the two sets of data on the same chart. Chart 4, however, will be different. With the reassessment data points added, Chart 4 will show you the difference between the baseline and the reassessment scores.

13.3 Analyzing the Survey results

The elements interact in two important ways. The first and less complex are the interactions among the elements in the baseline case: the initial survey. The second and more complex are the interactions among the elements between two sets of data. The charts in Figures 13-2 through 13-5 have only one set of data. They are therefore, easier to analyze. As you look at them, consider the following four questions.

Chapter 13

1. How well did you score each element?
The charts clearly provide this important information. If you take the maximum score of 20 and divide by 4, you get the five bands on the web of change. Starting from the center, the bands represent very poor, poor, average, good, and very good. The distance of the data point from the center in Chart 1 or height of the bars in Chart 2 give you some perspective about how you scored your company's performance.

2. How close does the data cluster around the average?
This information can be obtained from Chart 3. You want to see how closely the scores are staying to the study average. You want the change process to proceed in all eight areas at relatively the same rate. One element jumping out in front or falling behind the other creates an imbalance that can undermine work in the other areas.

3. Which elements differ significantly from the average?
This data is available from Chart 4. Although it is similar to the line graph in Chart 3, it shows more clearly which elements have large deviations, either positive or negative, from the average score. The deviations below the zero line indicate the elements needing improvement. The larger the negative deviation the more you need to determine what problems are in your system.

4. Do you believe the survey results accurately represent your gut feeling about your effort?
This question needs to be asked because you may get scores that just don't seem correct. The reason a score may not feel right may be related to the person who conducted the survey. It may also be related to a misunderstanding of a question. In either case, if you sense that something is wrong with the results, then question those who conducted the survey. Questioning the survey does not mean challenging the people who took the survey to change their scores. If you do this, the value of the effort will be lost.

Answers to these questions should open the door to analyzing what adjustments you need to make for your change effort to be successful. Taking the survey without analyzing the results, and then trying to make corrections, is simply a useless exercise. Learn from the results.

Chapter 13

As already noted, the second type of interaction involves two sets of data. One set of data is generated in the baseline survey. The second set is generated when you take the survey again after you have made a change within your organization. This second survey can most likely be completed after six months. I would not recommend a shorter time; remember that change takes time. If you repeat the survey too soon, the results may not be better. They may be the same or they may even be worse. Such results will lead to frustration. You could even find yourself abandoning an effort that could have added real value if you had only waited for it to take hold.

On the other hand, you shouldn't wait too long to repeat the survey. Waiting an extended period of time adds the risk that you could miss a problem that is brewing. Obviously you don't want this to happen. In some cases, the change may have been taking hold, but by waiting too long to analyze your progress, you allowed problems to develop that caused the change to fail. Therefore, waiting much longer than six months is not ideal either. Although I'm not casting six months in stone, use it as a guideline for when to repeat the survey.

Once you have taken the survey again, the questions you need to ask yourself are a little different than before. Now you are looking at multiple sets of data. The interactions among the elements are far more complex.

1. How did the scores change over time? Did the scores, either collectively or individually get better or worse, or are the results mixed?
2. Where are the large changes? Why did they occur?
3. Which changes did you expect? Which changes surprised you?
4. Which elements differ greatly from the average, whether by a positive or negative margin?
5. Do you believe the new survey accurately represents your gut feeling about your change effort?

As you can see, these questions are, not surprisingly, more involved. Trying to analyze one set of data is already a difficult task. Having two sets of data, with interactions not only among the elements, but within the same element over time, is even more difficult.

To help you with this analysis, I will present a simple baseline survey, then a more complex example that includes a second survey. However, let's first look at how to analyze the survey results.

When you see a low score you should ask "Why is this element (or group of elements) a problem." The operative word is why. Each time you answer the question, you need to ask it again. This approach will enable you to drill down into the details of any problem. For example, suppose that Technology has a very low score. You start be asking "Why is the Technology score so low?" Your answer may be that the software support tools that you have are not state of the art and are difficult to use. Stopping your analysis here, however, will leave you short of the actual reason for the problem, and hence the low score. You next need to ask the question, "Why is the technology not state of the art?" Here, the answer here may be that funding has never been provided to update it. Now you are getting somewhere, but don't stop yet. Your next question is, "Why hasn't funding been provided?" In this case, the answer may be that there has never been a single point technology contact in your company advocating improved software tools. You're closing in on the real problem, but you're still not there. The next question is, "Why has there never been a single point technology contact employee in this role?" The answer may be that the company, while recognizing the need, has never filled this position. In turn, you ask, "Why haven't they filled the position?" The answer may be that the company has received many resumes but simply has not had filling this position high on the priority list of things to do.

You can see by the sample analysis that asking "why" for each answer helps you arrive at the real problem: the company has not prioritized filling an important software position. You needed to ask "why" five times to get the answer. If you use this process, you will see that in order to zoom in on the real problem, you may need to ask "why" four or five times.

This type of analysis has another benefit. By correcting the root cause of the problem, you also solve the related problems. Thus, if the company hires a single point technology contact, then there will be an advocate for improved software tools, leading to increased funding and, ultimately, state of the art software.

Therefore, asking "why" repeatedly as you drive towards the root cause of the problem is an excellent way to begin solving the problem; but you still are not done yet. What do you do about the relationships between the elements? These also need to be addressed. I suggest you attack this issue in two steps. First, for each element, ask "does my problem with element A have any impact on element B?" Once you have identified the other elements that are impacted, you follow the questioning process for each. In the example above, you will undoubtedly see an impact between technology and work process, and maybe a few others. By following the "why" process for each of these, you will find more problems. However, you will also find more solutions that, once put into place will improve your effort.

13.4 Baseline Example

For the baseline example, consider a company that manufactures a single product throughout the United States. When the firm was originally established, it built similar plants in five key geographical areas in order to service the entire market with minimal transportation costs. Over time, the company has come to the conclusion that it does not want to continue marketing in the geographical area supplied by one of the plants. While the firm is not quite ready to close that plant's doors, it has steadily been diverting capital and resources elsewhere. The result is that the plant is old and beginning to show its age. Reliability has declined and with it, the morale of the employees. Many have left anticipating closure, though others have stayed, trying to make the plant successful.

Then suddenly an alternate use is discovered for the product, one that will be highly profitable to the company. The demand is so great that the company can not make enough. Therefore, the old plant is no longer on the verge of closure, but a new business strategy must be developed.

Element	Score
Leadership	5
Work Process	4
Structure	8
Group Learning	0
Technology	5
Communication	4
Interrelationships	5
Rewards	5
Average Score	4.5

Figure 13-6 Baseline Survey

Chapter 13

You are sent to the plant to turn the place around. The company needs it to become profitable and productive. Because you are familiar with the concepts of this book, when you arrive at the plant, you are ready to conduct the baseline survey. The results are shown in Figure 13-6 with the accompanying charts in Figures 13-7 through 13-10.

As you can see, the results are dismal. Nevertheless, the plant must be turned around and restored the high level of productivity that existed before. The first thing you do with the results of the survey is to gather your key plant personnel together to analyze your current situation. Taking each element and asking "why" repeatedly reveals many things that you can begin to address in order to make improvements. Your priorities lead you to the following results over the first six months:

- Bring revitalized leadership to the plant
- Redesign the work process and organizational structure to improve both work flow and employee productivity
- Improve communications so that employees understand what changes are happening and are able to feel part of the effort
- Improve relationships, even though employees are skeptical about the company's real motives
- Start a process to upgrade the computer systems that support the work

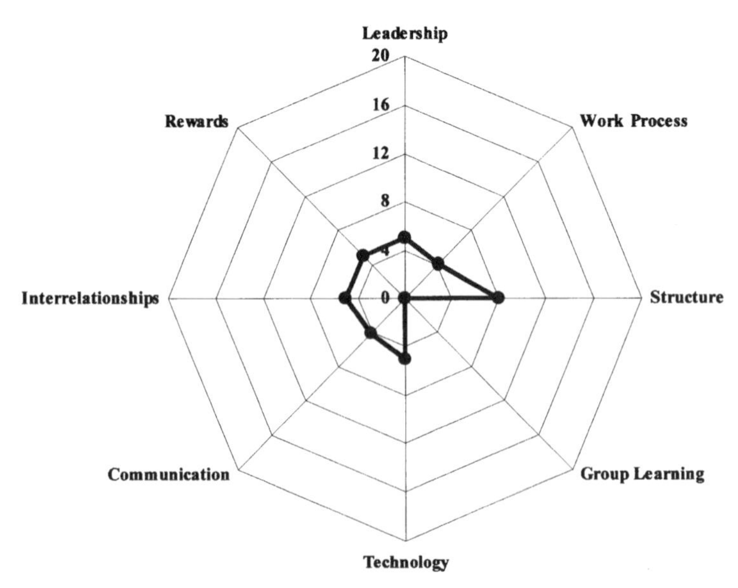

Figure 13-7 The Baseline Web

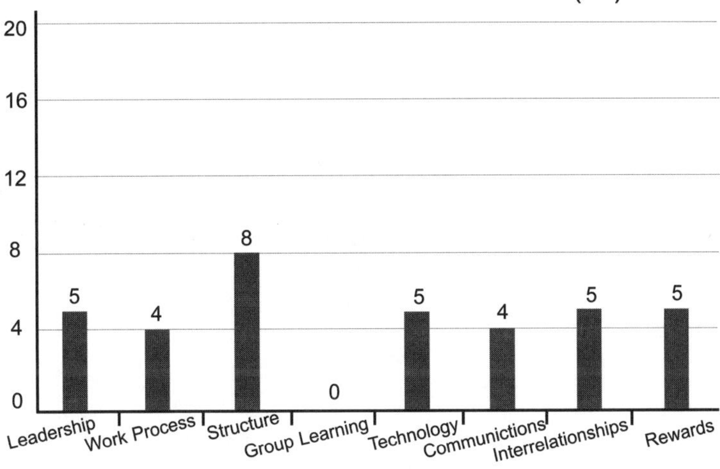

Figure 13-8

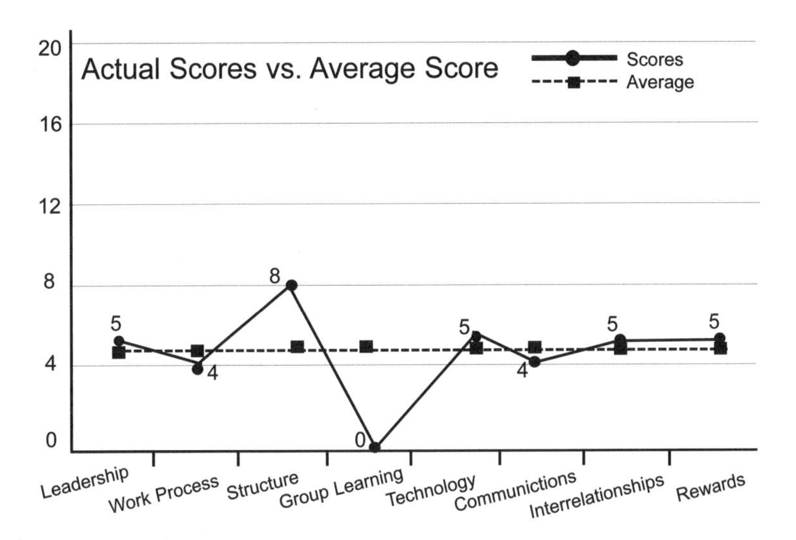

Figure 13-9

After six months, conditions are beginning to look somewhat better than that horrible first day on the job. Having read this book, you know that a reassessment after six months is a recommended. You're also confident that the site is ready for that review, based on all that has been accomplished.

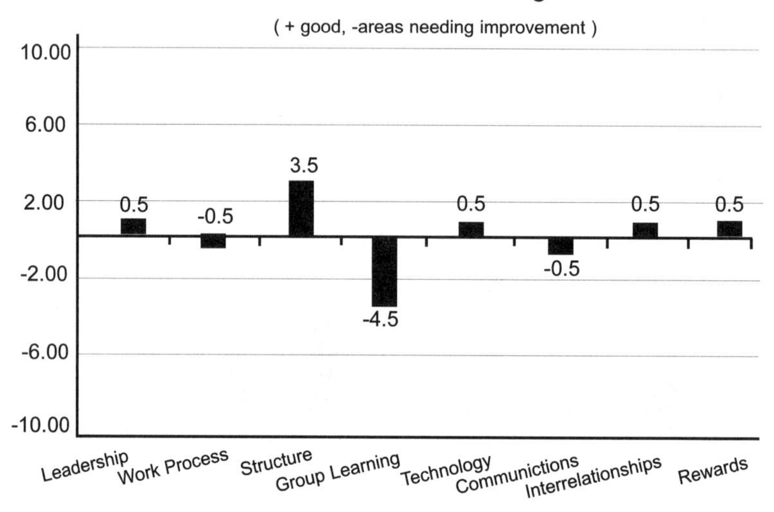

Figure 13-10

13.5 Reassessment Example

You decide to resurvey conditions at the end of your first six months on the job. Even though you hope that everything will have improved, you realistically know that this result is probably not going to appear. Not everyone at the plant has accepted your new ideas with open arms. Senior management is also having trouble with some of your ideas as well as with the way you have given direction. Because the plant was in trouble, you often had to mandate changes—this approach wasn't always accepted too well. The union is also having problems because you are trying to change some of the work rules in order to generate increased productivity.

In the area of technology and computerized support tools, the plant was in the dark ages. Improvements of this magnitude require a great deal of capital and time to execute. The technical area will get better, however, six months is not enough time to see these results.

You conduct a new survey and arrive at the results shown in Figure 13-11. The accompanying charts are in Figs. 13-12 through 13-15.

Based on the changes you have instituted, the improvements shown in Figure 13-11 are understandable.

Element	Baseline	Re-Survey	Change
Leadership	5	7	+2
Work Process	4	15	+11
Structure	8	13	+5
Group Learning	0	5	+5
Technology	5	5	0
Communication	4	12	+8
Interrelationships	5	6	+1
Rewards	5	2	- 3
Average Score	4.5	8.1	+ 3.6

Figure 13-11 Reassessment Showing Both Sets of Scores

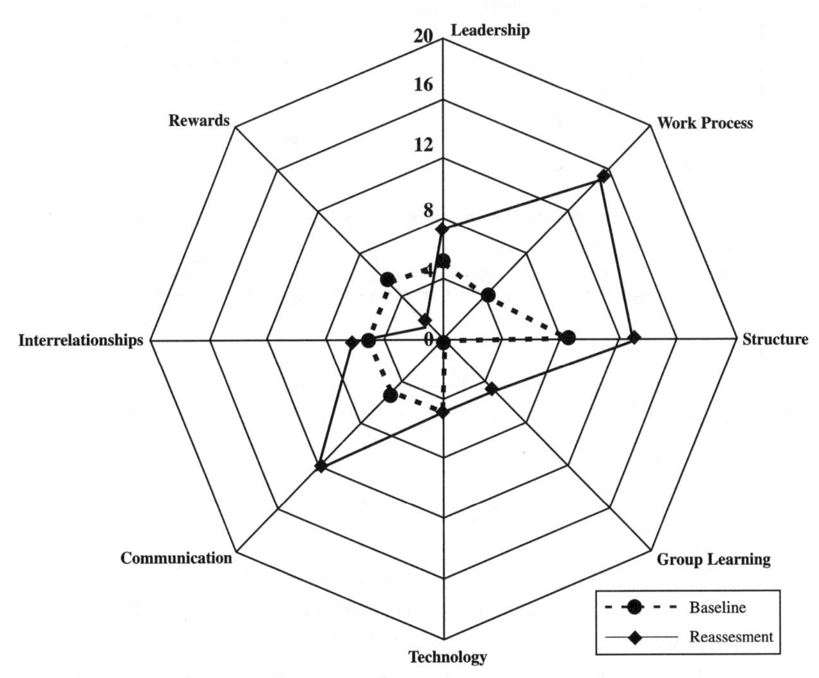

Figure 13-12 The Web of Change Over Time

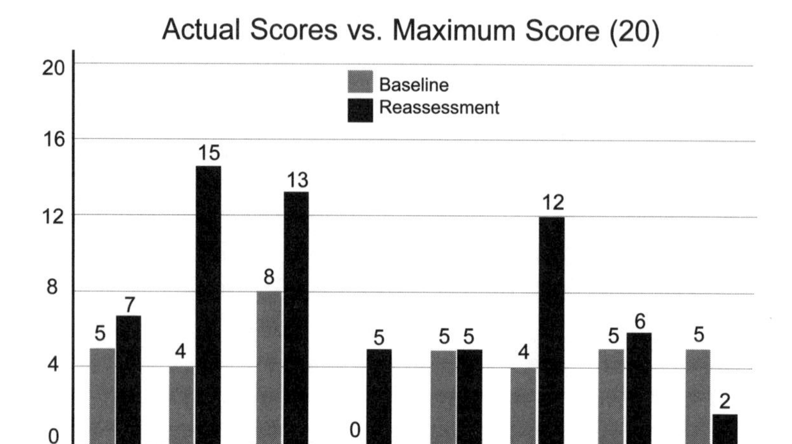

Figure 13-13

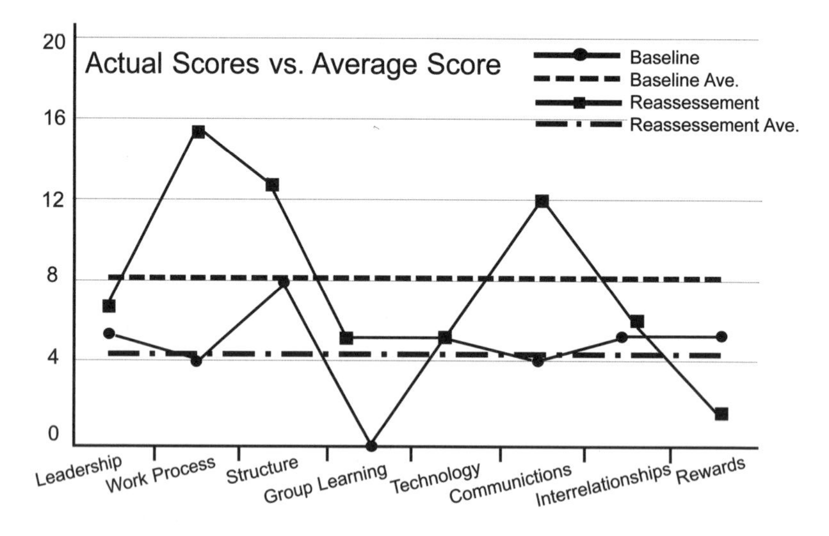

Figure 13-14

- Leadership improved by 2 points. Employees have hope for their future and the new leadership. However, they remain skeptical about the company's motives. Hence we don't see a large improvement. Figure 13-14 reinforces awareness of this issue. While the score has improved, it is now lagging behind the average of

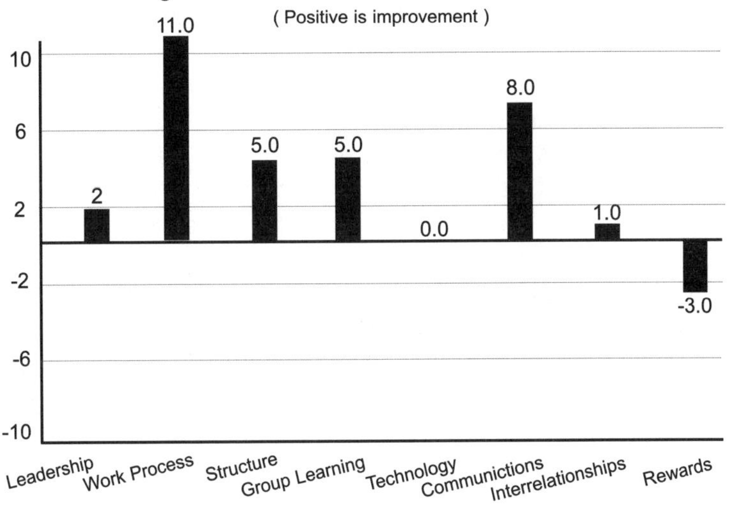

Figure 13-15

the reassessment scores. Therefore even with the improvement, a problem is emerging.Leadership is a critical component of a successful change effort. Therefore this problem needs to be analyzed and addressed

- Work Process improved by 11 points. This element has the most significant improvement, as should be expected. Without a good work process, the plant can not hope to become productive and meet the new demands of the market.
- Structure improved by 5 points. Although not as significant, an improvement as in the work process area, structural change was required to support the changes. This is an example of interaction between elements.
- Group Learning improved 5 points. Considering that there was none prior to your arrival, this improvement may be a result of your leadership style or other, yet-to-be determined factors. This area is another place for "why" analysis.
- Technology did not change. In fact, it has gotten worse relative to the other ele ments because it has slipped below the reassessment average score. This is represented in Figure 13-14. The most obvious reason is that improvement of this sort takes time, usually more than six months. However, other elements that can change more quickly have done so. Analysis is needed to make certain that this area is being adequately addressed. Significant improvement should start to show up by the next survey.

- Communication improved by 8 points. This significant change may not actually be due to improved communication. Employees may simply feel better about their situation, considering even a little communication to be superior to what it was in the past. The "why" process should be applied to determine the real reason before any action is taken.
- Interrelationships improved only 1 point. Even though the change is slight, the reason behind it is important. Relationships, both formal and informal, are a key success factor to a profitable business. This area is one where you must focus your efforts.
- Rewards declined 3 points between the baseline and the reassessment. Figure 13-14 shows that this decline was even more significant if you compare the score to the reassessment average. Because the plant has made significant improvement, something is wrong. A detailed analysis is needed and quickly. The decline may simply be a problem with the reward structure. However, it may go deeper, reflecting significant problems on the horizon, such as employee unrest or high turnover to come.

The example above is purely hypothetical. Still, I hope it helps to show the value of the survey, the reassessment, and most important the analysis process. These will help you determine the problems associated with the change initiative that you are putting into place. The data by itself, though important is not sufficient. The analysis and the action that you take to correct the identified problems is where the value lies. Correcting these issues early will reduce problems and support the change initiative as it moves forward. Not doing the analysis and simply attempting to correct concerns without understanding why will not make them go away. On the contrary, problems will simmer below the surface. They will continually worsen until you cannot correct them without major obstacles and delays to your effort. Therefore, use the survey and the web. Gather the data in the appropriate time frame. Most of all, complete the analysis and take timely corrective action.

Chapter 14
How to Handle the Process of Change

14.1 Keep Your Vision in Sight

Up to this point we have discussed preparing for change and using the Web to measure progress based on the eight elements of the change process. We can now look at the actual work associated with change. As you work your way through change, never lose sight of your original vision. After all, it represents where you and your organization are trying to go. The road to your ultimate goal will be neither straight nor smooth. In fact, as you go through the learning spirals, you are likely to head along paths that you never even thought about when you started. Maintain your focus. Use the tools that you have to assist you: The Goal Achievement Model, the Roadmap, and the Web of Change.

14.2 Types of Change

Recall from Chapter 2 that change can be categorized as incremental or step. Incremental change is easier; it requires less work and needs fewer people involved. The reason for this is that incremental change consists of small adjustments to the larger work process. The effort and its impacts usually affect only those immediately involved. Suppose you have a routine maintenance work order process and you want to add the ability to track emergencies. You can make the adjustment to the work order process with the involvement of relatively few people. In addition, any training efforts are simple because only a small number of people are affected.

Step or radical change is much more involved. Large step changes affect, at the minimum, entire departments and often entire plants or companies. Several years ago, I was involved in a radical change effort. We decided to install a common maintenance and materials management computer system in all of our plants. This effort required that we first analyze all of the existing systems and databases. Since each site was giving up its old system, we needed to determine how to convert each one into the new common system. Needless to say, this change was a major project, one that required a lot of time and money.

This chapter will focus on step change. Although incremental change is important, the real hard work is in the arena of step change. Furthermore, everything you use to tackle step change can easily be used for incremental change as well.

As you move through the change process, keep in mind the old saying that things usually get worse before they get better. When you start, you will probably see a lot of initial excitement and energy. Very shortly

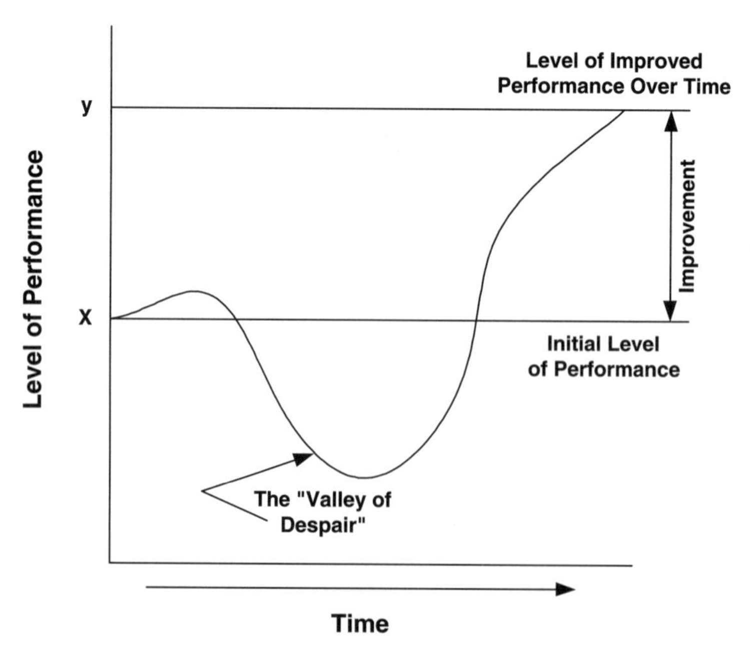

Figure 14-1 Performance Levels Over Time

thereafter, things slow down. You might then think the effort has failed or you did something wrong that has caused you not to achieve your goal. The diagram in Figure 14-1 illustrates a typical progression through the change process.

The y-axis represents performance level. The letter A marks the level of performance at the outset of the change initiative. The letter B marks the level of performance you expect to achieve once the change initiative has been implemented. The x-axis is time. I have not shown a specific time scale because change efforts are different, each requiring a different amount of time. You can see an immediate jump in performance when the effort first begins. This jump usually results from the initial excitement generated by the rollout of the new process. Soon after, performance usually declines below the starting level of performance. This decline may result from the typical difficulties associated with a large change effort. However, if you have prepared the change effort well, and you can stay focused on your plan, performance will ultimately improve, far exceeding the level at which you started. I often refer to the bottom of the trough as the "valley of despair". If you have ever been through a major change effort, you will recognize the feeling you had when you were in the valley. You will also recognize the positive feeling you had when you remained focused and worked your way out.

As you prepare for the change effort, you should also consider what you plan to do with the old way of doing business. Some companies that change, leave the old process in place; in a sense, they run two parallel processes until they get the bugs out of the new. In most cases, this is a bad way of proceeding through a change initiative. By running the old and new processes in parallel, you are providing employees the ability to stay with the old process. Running parallel processes will delay a successful change effort. It can even destroy it.

So what do you do? You can destroy the old process or render it into a state that it can not be used by others. They might be able to reference it for information, but cannot substitute it for the new process. Occasionally you can retire the old process or equipment so that it can only be brought back by a deliberate decision made by those leading the change effort. The old system then becomes a temporary back-up system to be used only in emergency circumstances. However, if you have prepared your change well, then the new way will be far better than what was left behind. The trick is to keep the process moving forward.

Chapter 14

14.3 How the Organization Should Be Involved

For major change efforts, a large portion of the organization needs to be involved. As the effort progresses, buy in is very important. Just as critical mass is important to teams, it is equally important for mobilizing the organization to embrace change. Involvement must come from top management and the client for the change effort. It must also come from those in the organization who will be affected as well as the project or change team itself.

Top Management

Although top management can also be the client, assume that the client group is one level below. How should top management be involved? First, they need to visibly support the effort. Change efforts of this nature do not happen in a vacuum—everyone knows some change is being planned. Top management plays a critical role by supporting the initiative and explaining why it is being put in place. Change is usually prompted by a business need or performance gap that needs to be closed. The reason for the change can be explained to the employees so that they understand why the effort is taking place. It is a serious flaw if top management doesn't provide this information and a level of assurance. Otherwise, the grapevine will create something to fill the void. Not only will the grapevine information be incorrect or incomplete, it will damage morale, slow down employee productivity, and make the change more difficult to implement.

Top management can also visibly support the effort by providing resources. Not all resources are in the form of money, although funding is very helpful. Often more important, and sometimes harder for an organization to provide, is the time of its people. If the top management visibly provides high quality employees as a resource for the effort, the organization will take notice and begin to understand the importance. Finally, top managers should periodically check in with the project team. This interest also sends a strong message to both the organization as a whole and the project team.

The Client

For this discussion, assume the client is the middle level of management within the firm. The client's involvement with the project team is essential. The client is, after all, the group for whom the change is being designed. They need to communicate continually with the project team, ensuring that what is delivered fits the task that the team was originally set up to accomplish. They can have a periodic review of the team's progress, evaluating where the project team stands relative to the work plan and deliverables defined in the scope. This communication must be a two-way effort. The team needs to make sure that its work is in line with the client's expectations; the client needs to do likewise. Failure to communicate can have serious consequences.

Several years ago I was asked by my boss to determine if our computer system could adequately handle cost control for one of our larger engineering projects. In this case, I was a project team of one. As I started to look into the situation, I discovered other problems with the system. Based on this discovery, I broadened my scope to address a much wider effort: why the cost control process wasn't working. When I finished, I proudly presented my work. The response was not what I anticipated. What I turned in was not what he had asked me to do and it wasn't what he wanted. Not only that, but I had angered many people within his organization for no valid reason. This experience provided a lesson that I never have forgotten: know your client and periodically validate your direction so that you know your work is on track.

The Organization

It is safe to say that if a major change initiative is being developed, the organization knows about it. They might not know all of the details, primarily because the team has not developed them, but they know something is in the wind. If the project is not intended to be confidential, such as planning for a massive layoff or relocating out of the geographical area, there are ways to involve the organization and develop their support. These ways include:

- A communication process where employees are periodically kept up-to-date on the effort
- A peer review process for team members to discuss some of the more important work efforts with their peers in order to get additional insight
- Bringing selected individuals from the organization into the process to generate their comments
- Having the project team create sub-teams of employees with the expertise to address specific issues

The bottom line: whenever you can provide the organization information and show the benefits, even before the change is put into place, the better it will be for the overall effort. Communication gives employees a chance to discuss and adapt to the new ideas before they become reality. Furthermore, they may provide you with insights that you can incorporate to make the final result even better.

The Change Team

Most efforts require some form of project or change team charged with the responsibility of investigating the current process and designing something new. Who becomes a member, how much time they are allotted, where and when they meet, and the actual process they follow are the subject of the balance of this chapter.

Chapter 7 discussed the actual team structure that needs to be in place in an organization to make change work. The change teams discussed here are somewhat different. They are groups of people pulled together with one single overriding purpose—to design in detail a change that will be implemented in the organization.

14.4 Defining the Approach

Before we can address the functioning of the change team, we must look at the two approaches they can take in defining what the change is going to look like. The first approach is to create an "as is" model. This model reflects what exists today. Once it is completed, you can then build the "to be" model. This model shows what you want the process to look like when you have achieved your goal. Use the vision and the plant goals to develop the "to be" model. The team can then identify the gaps between the two models, recommending the changes necessary to bridge them.

The second approach ignores the current process. Instead, you start with a blank sheet of paper or an empty board. Create a process that reflects your vision. Do not be confined to the process used before. Because the starting point is so different, this approach usually leads to a very different result than the first approach. It allows more creativity. However, it is more difficult and time-consuming. In the end, you still need to identify the gaps and make changes to eliminate them. The difference is that the first approach is anchored in the past, the second approach in the future.

How you decide which approach to use often lies in the degree of change that you think is required. If the existing process essentially works, but needs some adjustment, whether minor or major, then start with the "as is" model. However, if the existing process is in need of major overhaul, or simply is not working, then starting with a clean slate may be a far better approach. The two approaches are described in Figure 14-2. Notice that in both cases, you reach a point where you identify and develop solutions to close the gaps.

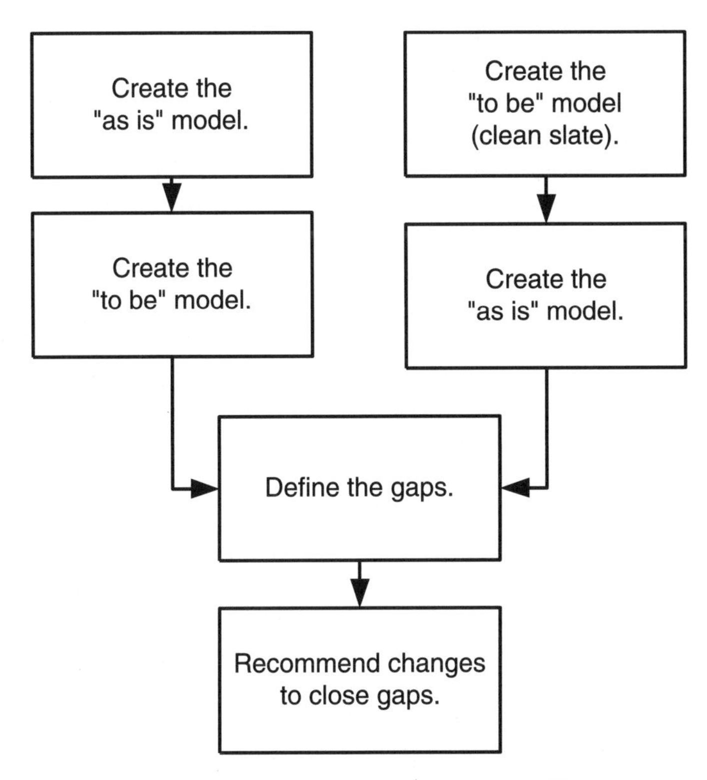

Figure 14-2 Methods to Redesign the Work

14.5 Change Teams

How you create the team that will design a change initiative and how you handle the logistics of their work effort are important factors contributing to the success or failure of the change effort.

Who

The definition of a team, introduced in Chapter 7's discussion of the process teams responsible for the day-to-day work, applies equally well to project teams. The difference is that the performance goals for which a project team is held accountable are a set of deliverables from the design project, not longer term process results.

One thing is for certain. Project teams are not assembled without a great deal of work identifying who will serve on the team. Not only do the members need complementary skills, but also they need the right skills—both functional and process. Depending on the type and size of the change, team members may also need to be at certain levels within the organization. This is not to say that employees at lower levels could not handle the work. They may be well qualified, but typically you need team members at specific organizational levels in order to assemble the right level and mix of skills. Team members generally need to be peers, rather than finding themselves in a manager-subordinate relationship. I have seen multilevel teams, but they are difficult to manage. The results are usually not satisfying to the members at the lower levels. All too often, the senior managers take over and monopolize the discussion. Those lower in the organization wonder why they were invited. The experience sours them for participation in the future and, in the end, lessens the potential value of what they could contribute.

Businesses that have an hourly workforce should, at times, include hourly workers on project teams. They too can add value to the process. Although the manager–subordinate relationship can get in the way, I have seen the approach work well in both unionized and nonunion plants. I have also seen it be a terrible mistake, and cause many problems. My advice to you in this area is be careful. If it seems like the relationship is such that value will come from the effort, by all means give it a try. If you are a little gun-shy of a full blown effort, consider asking a group of hourly employees to review your ideas after they have begun to take form. You could approach them simply by asking them for their input. If they are interested, they will meet with you and provide the input you seek. Otherwise, you have at least given them the opportunity.

As for complementary skills, functional knowledge is not the only requirement for a project team. You also need team members who are open to new ideas, and willing to think outside of their day-to-day experiences. Assembling this group is not easy. These employees are the ones usually sought for other similar efforts. Yet if you create a team with members who do not have these skills, problems will result. Furthermore, the process will take a lot longer. You will continually have to overcome resistance to change from those team members who are anchored in the present.

The departments where these quality employees work may not want to give them up. It is equally true, however, that in order for the change initiative to be successful, they must. Although you would hope that this fact is clear to the senior managers of the various departments from which you want to draw, this is not always the case. They may understand the broad perspective of what you are doing, yet they may not see the specific connection between the employees you want and the end product. In this case, you must meet with these managers to convince them of the need. Point out to them that the employees they provide to the project team also indicates their own level of commitment to the change process. Because these senior managers are probably committed to the effort, this point, along with a meeting to review the scope of the project, may help your cause.

Finally recall that facilitation is important for the group's success. You don't necessarily need a true process facilitator (one who knows nothing about your content). Instead you may want someone who actually does know something about the content, but is equally skilled at running a project group so that the group can achieve its goals. The facilitator can do a lot to keep the team focused and the work moving forward.

What

Change teams must be focused on what changes you plan to make. At the time you are forming your team, the change initiative is generally understood, but the details have not been fully developed. Understand that because the details are not yet worked out, the end result of the change effort may not look exactly like what you or the team originally had in mind. Even with a detailed plan, spiral learning will lead to adjustments. As the project team goes through the process of developing the details, it will learn things that will alter the final outcome from what was originally conceived.

You also need to develop a work scope. Clients have expectations of you as the leader and of the group conducting the redesign. You want to deliver the best possible solution. Therefore, you need a scope that defines what you are going to do and what you are going to deliver. Once this scope is completed, you can make sure that your understanding of the clients expectations is clear. If you are in agreement, then you can move forward. Otherwise, you need to change the scope to reflect what is required.

The scope has other purposes as well. It helps you to measure progress against an agreed upon framework. It also helps discuss changes in the event you identify additional work that will provide increased value for the project and the client. The scope should be given to the team members before your first meeting with them so that they know what to expect. In turn, you will use the scope to manage the work effort.

The following list details information that should be in the scope? You and the client may want additional information; by all means include it. The more definitive the scope becomes, the more assured you can be that you and your team are in alignment with the client's requirements. The scope should include:

- A cover page to distinguish the scope from the general memos and letters circulating through your firm. When people see the scope, you want them to know it is different and take notice.
- A mission statement so that the purpose of the effort is clear to all who read it.
- Goals associated with the overall purpose.
- Logistics including how long the effort will take, meeting times, and locations.
- Team members (by name and work group).
- A statement of team guidelines—what the team can and cannot do. For example, a guideline may be to avoid changes that would exceed a specific budget.
- A project timeline to expand upon the overall timeframe of the effort as well as a list of milestones that you and the clients can use to track progress.
- A description of the work process you will be using. These work steps should match up with the project milestones.
- An explanation of any cost associated with the work. For example, if you plan to meet off site, list costs such as the meeting room, food, and transportation. You should also include internal costs, specifically, the time of the team members while they work on the project.
- A set of deliverables from the effort, including an explanation of whether the results will be transmitted in a presentation, a report, or some other format.
- A list of benefits that the client will receive because of the approach itself, separate from the final results

WHERE

Where your team works is very important. Most organizations prefer that these meetings be held onsite to save on expenses such as the room, food, and supplies. However, if the effort is significant enough, then there may be value in taking the group offsite. In many cases, especially for important initiatives that are not massive in scope, a local hotel will suffice, even just to get people out of their day-to-day environment. What is the benefit of relocating the work effort? First, team members feel special; they seldom get outside of the plant during regular work hours. Second, their minds are cleared for the difficult work you are asking them to do. When you are literally outside of the plant, you can often think more clearly about what occurs inside the plant.

Insist that team members turn off their pagers, cell phones, and radios. Make a rule as well that the members cannot run to the phones at breaks to listen to voice mail. There is no quicker way to lose continuity and the group's creative juices than to allow access to the electronic devices that reconnect them back to work.

When

The timing of the effort is also important. Many people think that you can undertake project work of this nature eight hours per day everyday for the duration. Sometimes, especially if you have a short timeframe in which to get the work done, this schedule may be necessary, but it is seldom productive. Creative work of this sort is very tiring. At the minimum, provide a 10-minute break every hour. Plan your meetings with the needed break points. Provide at least an hour for lunch so that people have a chance to relax. However, keep the group together during lunch. It is amazing what can get accomplished in an informal environment.

Another important aspect of timing is scheduling the meeting content. I have facilitated many of these sessions and have found that the most productive thinking is in the morning. People are fresh and rejuvenated from the previous day's work. In the afternoon, many start to get tired. If any of them have consumed large quantities of sugar at lunch, you may even lose them for an afternoon nap. The best strategy is to schedule the thinking components in the morning and any presentations, recaps, or discussions in the afternoon. Furthermore, don't be afraid to quit when you feel the group has run out of energy. I have been at sessions where the manager felt it was necessary to "work until we are done." I sat there waiting for the meeting to be over, not contributing. It wasn't that I didn't want to contribute; I was just tired. It would have been easier and more productive if we had waited until the next day to finish. I have also participated in meetings held after work with the same negative results. My recommendation is to start early and work hard in the morning when you can get maximum value from the group.

Chapter 14

How

There are as many ways to run a project as can be imagined. Over the years, however, I have found that several key components are always part of a successful process. These include:

- Provide the scope of work to the change team in advance of you first meeting. They then have time to think about what they will be doing. This approach is better than handing out the scope the first day the group assembles.
- Explain to the team members why they are there and why they were selected. This explanation provides immediate positive reinforcement —they were chosen because of the value that they could provide.
- Have everyone in the room state who they are and their credentials (such as current and past jobs and educational background). Sharing this information puts everyone on an equal footing, especially those who know few, if any, of the other team members.
- Review the scope so that members can discuss it or ask questions. Because you distributed the scope beforehand, they should have read it and be prepared with questions. This discussion begins the team development process. At this point, any scope changes or additions should be addressed.
- Define the work effort. Give everyone a common understanding from your perspective as the team leader (who has already gone through a similar process with the client). Spend any necessary time working on the definition before you move on. Once complete, it assures that everyone is on the same page for the balance of the effort.
- Define your approach. This discussion was already addressed in detail in the previous section.
- After defining the approach, the team needs to perform the work for which it was created. The scope provides the team with the mission and the goals, but it is up to them to determine the initiatives and activities. This work is often difficult. Members of the team often have different ways of viewing the same issue. Take the time necessary to reach consensus. The results from a process where the group developed recommendations driven by consensus can be outstanding. Furthermore, the team will feel good about what they accomplished.
- The last part of the process is the presentation, when the team makes its recommendations to management. Before this presentation takes place, the team should understand and expect that some of its recommendations will not be accepted. Such expectations should not interfere with any feeling that members have about the quality of their effort. If they are able to get even a small portion of their recommendations accepted and implemented, they will have made a significant contribution. It is even possible that after the first changes begin to take effect, the organization may undergo some spiral of its own and implement the rest.

Chapter 14

14.6 More Keys to team Success

This chapter has looked at many of the components and processes that go into making a project team successful. As the team develops and implements a new process for the organization, it needs its own process by which to work. This last section introduces some additional tips about how to make the team's process successful. Most of these stem from my own experience. You may have your own tips to add to the list.

The information stays in the room. Until approved by your client, the work you do is neither final nor official. Regardless of the approach you take, ideas will be generated that, if released to the general population, could cause problems. Simple ideas such as a restructuring of an individual work group can cause serious problems if it is leaked early. Furthermore, when the idea is presented to management, it may be rejected as a solution. You don't want employees spending their time and energy reacting to something that will not happen.

Time is needed to allow the team to do the work. All too often, the client places too short of a time constraint on the effort. The constraint is too short not only from the standpoint of having to work hard, but also for allowing the team to form and complete its work. A cycle of team formation (e.g., moving from a group of individuals to a team) must take place in order for its work output to be valuable. If the time allotted is too short, this cycle will not take place.

Recognize early obstacles. A team is usually made up of employees with varied skills from varied functions within the organization. The members each come to the process with a different set of ideas as to why they are there. No matter how many times you tell them ahead of time or at the start of the meeting, the group needs to collectively reach this understanding. Once past this early obstacle, the group will be able to work as a single unit. If you are running the session, watch for it and allow the group time to get past it. Sometimes this means stopping the team's work to specifically address the issue.

Allow war stories, but not too many. In many teams, members spend a lot of time telling war stories—personal stories about things that happened to them that seems relevant to the issue being discussed. Many people who facilitate sessions try to cut these stories off under the belief that they don't add value to the work at hand. This is not true. Even if they are not directly tied to the work, sharing personal experiences helps to bind the individuals in the room into the team. Allow these stores to unfold. Encourage them to be told during breaks. At the same time, don't allow so many so that the work suffers.

Don't let the group stay in "they won't let us" mode. A lot of people frustrated by the current process believe that the organization will not allow change. If this attitude becomes a central theme of the work then the team will not deliver anything of value. The group was empowered to recommend change. It needs to approach the effort with that spirit. Even though some recommendations may be rejected, the group needs to recommend what it believes is right. Pay careful attention and look for this attitude. When you see it, make it stop. One approach is to draw a line down the middle of a page from a flip chart. On the left column write "we can't" and on the right column "we can". Tell the group to complete the chart as a team listing all of the reasons for both categories. When the chart is completed, tear off the left "we can't" side and throw it away. Explain to the members that the reason they are all there is to show how change can be accomplished. This exercise may seem simple, but it generally makes the point.

Complete important offline work. During the day, the group will complete a large amount of work. A lot will be written on flip charts and note pads. If you are running the sessions, type up the group's work in an organized fashion after hours. Distribute the summary to the team the first thing the next day. This follow-up work involves a lot of effort, but it is important to the team's success. Having a clean and organized hard copy of its efforts from the previous day helps the team focus on what was done and move forward.

Keep the group on track, but don't be rigid. A rigid approach leads to rigid thinking. The team needs to be controlled, but not in a way that breaks down its ability to think about new and different ways to work. Especially when a new skill is being learned, flexibility is required.

Start and finish on time. Too many companies call meetings and are willing to accept lateness. Team members need to show up on time. Similarly, the meeting needs to start and end on time. Enforcing these hours shows members that their attendance is important; it serves to solidify the group. Late arrivals are disruptive, especially if you have to take time to tell them what they missed. If members are going to be late, or have conflicts due to previous commitments, you need to tell them what they missed outside of the regular meeting time.

Chapter 15
Measurement

15.1 What Gets Measured Gets Done

It is not enough just to set up and execute a work-process change. You also need tools that will sustain the effort into the future. That is the role of measurement. Virtually everyone working in any industry has had experience in the area of measurement. You have probably used measures to track your work results, goals that you promised to meet, and tasks that your boss wanted done. You have experienced measurement being used both positively and negatively. This chapter explores how measures can support and sustain your work change process.

Whatever gets measured by an organization receives the majority of its attention. Simply by virtue of obtaining and displaying data, you and your organization are focusing, at least on a minimal level, on those areas that you are measuring. If management is astute, you will be asked numerous questions as to why you are or are not conforming to the plan. If your measures are not tracking as expected, corrective actions usually follow close at hand. In reality, therefore, whatever you and your organization decide to measure sets up a subprocess that ensures more attention is given to these areas vs. those things which are not measured.

Measurement is usually performed on a periodic basis. A process needs to be set up to gather the data. Sometimes many employees, work many hours, both within and among departments to obtain what is needed. This information is then usually reported in some common communication forum, further indicating to everyone that the measure is important. This cycle is self-reinforcing. Conducting the measurement places even more focus onto the measures, compelling even more effort to make them successful.

Although some measures add value, others have little or no value to our organization. These measures are requested for many reasons. Some may check what is taking place in the work process. Others may result from a unilateral decision somewhere in the management structure, or be conducted simply because they have been in place for a long time, even though no one realizes that its original intent no longer exists. Still others are in place because someone with a good intent selected a wrong measure. These measures also get handled by the organization at the compliance level. They may keep the boss happy, even if they do not really add value to the work—in our case, to the process of change.

Yet the truly valuable measures bring focus to the organization. These measures promote understanding, clarifying what is working or not working, and even what is working well, but could still use improvement. These measures need to be continued.

The statement "What gets measured gets done" needs a corollary. It should state, "What gets measured gets done, but you need the right measures to get the right things done." This raises an important question: What are the right measures?

15.2 What Should We Measure?

What should be measured is different for every company. Because each company, or even plants and offices within the same company, are at different stages of change, the appropriate measures may be completely different or at different levels of sophistication. So there needs to be some thought given to the selection of these measurement tools.

Select and use measures to influence behavior. For example, measuring profitability will positively influence behavior. If your firm was unprofitable and your change efforts turned it around, a chart measuring the effect on profitability would show employees that their efforts had been successful. As they watch the chart reflect the change from an unprofitable to a profitable operation, the measurement reinforces their hard work to improve performance.

Chapter 15

Influencing behavior in a positive manner is not the only possible result of measurement. Some measures have the opposite effect. Suppose that you were downsizing as part of a restructuring effort. You maintained a chart that measured head count—the number of people remaining in the organization—as the department undergoing change reduced its size. This information displayed on a measurement chart could have a negative impact on those groups that had not yet reduced staff. Their natural interpretation would be that management was only interested in layoffs—why else would head count be featured as a measure? In turn, their main concern would be focused on protecting their jobs. With this type of self-defensive thinking, these groups would support little in the form of a change process. It may, in fact be the case that other groups were not targeted for downsizing. The use of head count as a measure, then, would have a negative impact.

Some measures, such as plant overtime, can influence behavior both positively and negatively. Suppose your company is showing a very high use of overtime, cutting into the bottom line profit. Supervisory positions are staffed by foremen who work overtime as part of their salaried job. For supervisors, this overtime is not considered extra work, and they receive no additional payment. On the other hand, the mechanics work overtime at time and one half for all of the hours they work. A work process change is put into place to reduce the overtime. The change will have positive effects on profitability. Furthermore, the foremen will be pleased because they will work fewer hours with no less pay. But what about the hourly mechanics? As overtime is reduced so is their potential to earn additional payment. In essence, they are losing money. Will they be happy about this change?

It is important to identify what you want to change in the work process, then carefully put measures into place that will affect the selected behavior in the right direction. The key word in this last statement is carefully. As you can see from the simple examples described, you can easily introduce problems if you use the wrong measures.

Be aware of the time needed to assemble data and generate measures to portray the information. Measurement takes people's time. Data needs to be gathered, sometimes from more than one source, and then summarized so that it can be understood. These efforts need to be completed in a timely manner. Therefore, establish a process for measurement that makes these gathering and summarizing steps routine and does not over burden the organization. In addition, don't continually change the reports and the reporting processes. When you do this, you cause those involved to lose energy. You also send a strong message that you don't really know what you want. This message can lead employees to conclude that you do not know what you are doing. Changing reporting processes is a great way to undermine your overall effort. Once you have lost your credibility, it is very hard and often impossible to get it back. Because measures influence behavior, this is not the type of influence that you want to exert.

15.3 Selecting the Measures

It is time to decide what you want to measure. You only want to go through this selection process one time—and correctly. Do not undermine the process by rushing the selection step. As you go through the process of selecting measures keep the following points in mind.

Use measures that are meaningful to others. Although it is fine to measure things that are important to you, these measures are probably not necessary to change your behavior. Instead, as you select the measures, focus on those measures that will have meaning for others who are receiving the data. Examples include company profitability, rate of production, and project costs. The measure should have meaning for the receivers. How much attention will others pay to a measure that they cannot connect with their work or with their end of the change process? Probably very little.

Include others but stay focused. This point is a double-edged sword. It is important when you select measures to include those who will be receiving the data. When others are part of the selection process, they have a sense of ownership. They are more likely to get involved with gathering and analyzing data, and ultimately changing the measure for the better. A side benefit is that the measuring process becomes automatic with those involved undertaking the measurement work itself as well as recommending process improvements.

There is one pitfall, however, that you should be aware of as you proceed. At the same time that you get others involved, do not let them water down the measurement effort. They will do this for many reasons, whether intentionally or not, including;

- Lack of understanding of the need for the measure
- Different concepts about the level of detail required
- Suspicion of the motives behind the measure
- Fear of punishment if the result does not match expectations
- Lack of interest
- Reluctance to take on additional work

Any one of these above reasons, if allowed to become a part of the selection process for the measures, has the potential of destroying its effectiveness. Suppose you want to measure production. You turn this request over to the group that is working on measure selection and development. You may want production rates on a weekly basis. However, if you have not effectively communicated the desire to the group members, they may not select production rate as a measure at all. Even if they do select it, they may not arrive at the measurement frequency you desire. As a result, you may have something less effective than you wanted.

The solution is for you to provide the project team with a clear set of guidelines describing the kinds of measures that you want to see. You don't necessarily care about the details of how the group develops the measure, as long as you get the information you need.

Thus, if you are interested in the production rate, you need to explain to the group why the measure is important to you. Next you need to provide guidelines for their work effort. In this case, you want a measure of the production rate over a weekly time period. Your guidelines then are as follows:

- Production will be a measurement.
- The measure will be completed weekly.
- Data that should be included in the production metric should include items selected by the group.

How the group gathers data, develops information, and delivers the measure is for them to handle. You have provided a set of guidelines to keep the effort from being reduced in scope or effectiveness. Not all guidelines are as specific as these. I have been involved in efforts where the latitude of the group was much larger than in this production example.

Use straightforward measures. Some measures are so complicated that only a select few with a great deal of training will have any idea of what they mean. These are measures you probably don't want to use. You probably have your own examples that fit this category. Measures should be easy to understand. They should not be either difficult or time consuming to prepare. There is no easier way to demotivate employees from using a measure than to require a lot of their time and energy to obtain the data. Furthermore, measures that are time consuming reduce employee productivity in other important areas.

Sometimes, however, the data simply is difficult and time consuming to obtain. If that measurement is truly important, try to find alternate ways of obtaining the data. In addition, provide training to those on the receiving end so that they will we able to understand the measure.

Make your measures timely. Suppose you conduct an activity this month. You see the financial results for the month after the books have closed, about two weeks into the second month, and then recognize that a change is needed. You make the change at the end of the second month and see the results of your corrective action sometime in the middle of the third month. If you are not able to implement the corrective action until the start of the third month, you won't see results until the middle of the fourth month. By using this monthly reporting strategy, it will take three to four months to identify and make a correction. This is unacceptable. You could go out of business in less time. You need to link the activity with the measure more closely in time. Course corrections can then be made when the measures are still fresh. My recommendation is to look at your critical work process measures weekly, if possible.

At first, this recommendation may seem to contradict the previous point about simplicity and ease of measurement. Weekly measures require additional work. However, consider this example about the monthly financial measure. If you are able to make weekly measurements and address the problems promptly, you can make twelve course corrections in the same time that you could make only one in the example. Timely measurement empowers the change process.

Some measures are by necessity monthly or quarterly ones. However, a weekly measure will more directly link the measure with the work. This strategy should receive close attention.

Take small steps. Although measures influence behavior in a large way, behavioral patterns and routines generally change slowly. For this reason, take small steps with measures aimed at changing the work process. You can then keep the work process aligned with what you are measuring and help everyone remain focused. Small steps enable the employees who are involved with the change to get used to what you are doing. As their behavior and the process evolve, the measures can be adapted.

For example, you might institute a measure to track which plant areas are on budget. Once area teams are comfortable measuring the bottom line against budget, you can expand the measures to look at individual line items. If you start with the line item measures, you might have problems because the participants are not ready. However, six months after you start measuring the bottom line, the participants should have evolved enough so that you can enhance the measurements.

Communicate the measurement. Measurement highlights a goal you hope to achieve. By themselves, however, measures have little value. If you want to change behavior with measures, you need to communicate this information on a regular basis. Hanging a wall chart is not nearly as effective as a real person communicating the information and, if needed, answering questions. The electronic world of e-mail, the Internet, answering machines, and voice mail all have their place and their value. However, if you really want to make an impact with the measures, have a real person make the presentation.

Modify or improve the measure, but stay focused. This point may at first seem to contradict the importance of defining what you want to measure and then staying the course. However, just as spiral learning allows you to adapt goals and initiatives, it also allows you to adapt the measure, if the focus remains constant. By making the improvement, all that you have really done is to upgrade what is being measured. This is usually in response to those involved exhibiting a higher level of understanding about the process and their role in the effort.

Feed it or kill it but don't let it starve to death. Measures must be maintained as part of the change process. However, if a measure is not contributing value, drop it, finding a replacement that will add value. Do not maintain a measure that has little value, especially when everyone knows it. Measuring them casts doubt on the leaders, saying either that the leaders are too proud to change or they look at measures so infrequently, they don't recognize those that should be eliminated. Either feeding or killing measures is acceptable. Starvation is not. Starvation occurs when the measure is left unattended and unused.

15.4 Why Measures Fail

Measurement helps move the process of change forward, but quite often the best process and measures fail. There are four likely reasons.

1. The measurement is incorrect.
2. The measurement is being used for the wrong purpose.
3. The measure is being manipulated to show positive results when there are none.
4. There is an indication of organizational resistance.

The first reason can be addressed and quickly corrected, simply by finding other measures that fit within the context of the change process.

You can use some of the recommendations in the previous section to help you find new measures.

The second reason criticizes how management is using the measures. It needs serious review. Incorrect use of measurement tools are usually related to the practice of punishment for poor performance. If behaviors of this type are permitted and are widespread, they can easily lead to resistance and, eventually process failure.

The third reason requires deep investigation into the mechanisms that are causing people to feel a need to manipulate the results. If you don't take corrective action, the process failure may be a complete one.

The fourth reason focuses on organizational or individual resistance. Recall from Chapter 9 the discussion of sabotage. When resisters cannot openly resist change, they work in a covert fashion to sabotage the effort. If change measures are failing and the first three reasons don't provide you with a reasonable explanation look to resistance as the culprit.

Suppose you set up a series of measures to look at year-to-date cost vs. budget. The data for one month is available the middle of the following month. Over time you notice that your attempts to correct problems are not meeting much success. By the time you get last month's data and take corrective action, too much time has passed. Your dynamic business has made your correction useless. In this scenario, the wrong measure was used. A better approach would be to shorten the time between measures so that more timely corrective action can be taken.

Now suppose you measure work performed on overtime, then punish your subordinates based on the results. Without the need for additional description, this example is clearly one of using a measures for the wrong purpose. If you continue to punish people for bad results, and if the overtime rate continues to be bad, those involved will distort the truth. Even worse, your staff will spend inordinate amounts of time developing defenses and explanations to avoid punishment. They will literally have no time left to correct the problem.

Next suppose your company has initiated a work process change in equipment maintenance. One measure looks at compliance with the weekly schedule, in other words, how many hours of the mechanic's time was charged to work that was on the schedule developed the previous week. Over time you see that the compliance rate is high. However, when you visit the work place, you see frustration with the process; the work execution is in a state of confusion. Something is wrong with the measure because it doesn't match the reality of the process. A detailed study reveals that the process of weekly scheduling does not exist. The schedules are actually being prepared one day ahead of time. Therefore, compliance appears high when it is not. In this example, when the measure was checked against reality, it became apparent that the process was not working as designed and the measure was being manipulated. The clearly pointed out that corrective action was needed.

Last is an example of resistance. Years ago, I was involved in a quality process, described in Chapter One. Management required mandatory meetings that no one liked. Few attended them. One of the measurements of the work process was meeting attendance. The requirement was for 100 percent attendance and the measure showed 100 percent meeting attendance. Don't assume, however, that the meetings were being held. It was widely know that management was not checking this measurement. However, they had mandated that meetings be held. The organization resisted what we found to be a rather ridiculous mandate. By an unspoken agreement across the organization, compliance was reported at the required level, but the meetings never took place. If meeting compliance had really been checked, the resistance would have been obvious.

15.5 Accuracy

Some measures of the change process will be statistically accurate, most will not. A problem occurs, however, when management expects all change management measures to be exact and they spend a great deal of time trying to achieve perfect accuracy. For those of you who are in this category, you need to change your approach. Change management measures indicate trends. They provide information that then leads you to seek more definitive information for analysis. This is the true purpose of measurement in the change arena. If you spend all of your time looking for accuracy, you will miss the boat. Consider the 80/20 rule. In this case, you only need about 20 percent to know what is going on in 80 percent of your work process change. Don't waste your time looking for complete accuracy. Look for and analyze the trends.

15.6 How to Make Measurement a Success

A key to successful measurement is the idea of focus. However, not everyone can focus on everything all of the time. Some teams and organizational groups focus on their assigned measures. But each measure needs someone in upper management to acknowledge that it is important. In essence, each measurement needs a champion. Measurement champions are not necessarily involved in the actual task of measuring. Instead, their function is to get upper management to support and focus on the measure. They can help the project group in cases when upper management intervention would smooth the road and help get the work accomplished. At times, and in the correct situation, champions may also present the data. By doing so, they help everyone recognize that the measure is important.

15.7 Who Should Measure

Questions often come up about who should do the measuring. Some suggest that it be given to a clerk or assistant to handle; others suggest that a separate organization do the measuring. Still others suggest that the measurement owners be the ones who measure and report. Measurement owners are the individuals or groups with responsibility for what the measure portrays. For example, the production department owns the measure of plant production.

I believe that those who own the measures should also be the ones who track the data and report the findings. Any changes are then linked with the owner; their efforts are focused when corrective action is needed. If someone other than the owner does the work, the owner can too easily disregard the data.

One plant where I worked measured specific cost data pertinent to the plant's production teams. These measures were made and reported on a weekly basis. At first, teams and the members saw little value in what was actually simple measurement work. Nevertheless, continued measurement was required. Over time the teams began to question the data and what they believed was wrong. They began to get a better understanding of what the data represented and how it related to their budget. In the end, the measurements became a non-issue and the teams took over ownership of their budgets—our original goal. The measure, the continued focus, the simplicity of extraction, and management's nonpunitive reaction for being over budget changed the behavior.

15.8 Measurement Communication

Measurements need to be communicated. There are many reasons for this. Among them are the need to inform those who have asked you to measure something about your progress, and to compare your group with others, internally or externally.

As you determine how the information should be reported, start with the question: Who is your audience? Based on the audience, measurement can be communicated in three major ways:

- Upward communication, or informing those higher up in the organization
- Lateral communication, or informing your peers either inside or outside of your work group
- Downward communication, informing those who work for you or possibly others in your organization.

Sometimes your audience is a mixture of the three, making your task more difficult.

Next, determine what it is that you want to accomplish with your presentation. Is the purpose of your communication to exchange information, to report on the status of change in your group, to seek approval for funding, or to gain permission for something you need? Each of these purposes requires a different level of detail for each audience because each has a different level of interest in your data. If it is for information exchange, those in other groups with similar data will have a high level of interest because the data is being compared. If the purpose is for informing, there may be a low level of interest. If you are seeking funding or permission, the level of interest will be very high among those who have the responsibility for approving your project, but low among others.

Based on the audience and what you are trying to accomplish, the presentation of your measurement data will be different each time. As you determine the makeup of your audience, ask yourself what you want from them, and ask them what they want from you. With this knowledge, you can successfully prepare your presentation.

In all cases, assume that you will be asked questions. Be prepared to answer them. A good way to prepare is to have someone you know review your presentation and ask you a wide range of questions. This much preparation will give you some idea of how to be ready for the real one.

You can also choose other ways to communicate your data. Some of these include newsletters, memos, letters, flyers, and bulletin board postings. The general problem with these forms is the fact that they do not provide direct, face-to-face communication. How many times can you remember reading a handout, talking with coworkers, and discovering that each of you got different meanings from the communication? This is the risk of indirect communication. You risk the possibility that the recipients will place varying interpretations on the data and misunderstand your purpose. When you're trying to change the way an organization works, this lack of focused, consistent communication has the potential to undermine the overall objective. At the minimum, it can cause a delay in the effort, forcing you to spend additional time correcting and then recommunicating the information.

Therefore, combine the use of direct and indirect communication. When you use written forms, make sure that are also fully discussed in an oral (direct) setting before they are distributed. Then you will have agreement before others are left to interpret the information for themselves. The written communication then reinforces what you have communicated. Recognize that further interpretation is always needed as each employee applies the information to his or her own particular area. Thus, additional communication will be required in order to keep everyone on track.

15.9 Measurement is a Critical Success Factor

By now, I hope that you have a good grasp of what to measure, how to measure, and most importantly, considerations which need to be given to the communication of the measurement information. The change process can be greatly aided and reinforced by good measurement and its communication to the workforce. The process can also be seriously set back by measures which are incorrect, incorrectly used or communicated poorly. The bottom line is to think through what you are going to measure, make sure the purpose is aligned with your objectives, communicate it for understanding, and use it correctly. If you do those things, measures will keep you on track and vividly support the change effort.

Part Four
Completion and a New Beginning

At this point, we have talked about how to get started with a change effort. We have looked at all of the aspects that need to be addressed before we actually begin to implement the change as well as ways to monitor, measure, and communicate progress. The last thing we need to address is what to do after the effort has been rolled out, once the workforce is engaged in the process of changing. Change is a continuous process. Once started, it never stops. Part 4 discusses techniques to assure that we never forget this very important point.

Chapter 16
Techniques of Continued Growth

16.1 Change Efforts Don't Survive on their Own

At some point the change effort that you have designed and planned has been implemented. Your tendency may be to congratulate yourself on a job well done, then sit back and relax from your arduous task. Too many people who have implemented change initiatives relax, believing that they have completed their work. Not so! You have just begun.

Why do people let down their guard in this manner? Part of the reason is the natural let down after a very intense work effort. However, that isn't all of it. The other part draws all the way back to the linear project model presented in Chapter 3. If employees are thinking in a linear fashion, then they believe that once the implementation is complete, so to is the work effort. To them, implementation marks the end of the linear project process. However, the model to follow for change initiatives is not the linear model, but the spiral model. The process of change never ends. You plan, you execute, you evaluate, you learn from what you have done, and then you plan the next phase. In addition, you learn from the process spirals already completed so that each phase builds on earlier ones.

Some of the best change initiatives have been implemented successfully only to fail over time. Were they flawed in their design or implementation? If you and your team have done a good job, the answer is usually no. The flaw comes from the belief that the process, once implemented, will take hold and be self sustaining for all time. Unfortunately, this is not the case. New processes need care and feeding over the long term so that they can thrive. The excitement that was part of the rollout fades. Without a plan to help the effort grow, work habits can tend to drift back to the old way of doing business, even if a process is in place that makes the return difficult.

Many years ago I helped in an effort to put a team structure in place at the plant where I worked. The process was well thought out. Detailed plans were created so that a successful change initiative would be the result of the effort. Management was bought into the process and we implemented the change successfully. Then, as many organizations tend to do, we put the process on autopilot. We believed it could continue to add increasing value to our company with no further effort on our part. Several months later, expecting the process to be working, we realized that both the process and the structure we had created were gone. The initiative had eroded to nothing. To the outside observer, the plant looked like change had never been introduced.

Many things had to go wrong for a healthy, viable team process to deteriorate. Some of these include:

- Senior management, believing that the new process was working fine, directed its attention to other issues. In turn, the teams developed the feeling that they were no longer important.
- Middle management—the group responsible for the team process—had difficulty resolving process issues. No one was available to help sort out these process problems as they arose. As a result, problems undermined the effort.
- Over time, as employees moved to new jobs or other responsibilities, the team structures changed. Unfortunately, a review process did not exist to identify these structural changes and take corrective action.
- The teams came up with interesting new ideas. However, no continuous improvement process was in place to help get the ideas implemented. The ideas were eventually abandoned.
- No communication process was implemented so that individual teams would have a sense of being part of a larger community. Each team felt alone, without support.

The change leaders did not prepare for life after the change so that the teams could grow and thrive. As a result, the teams withered and died. Our own lesson was that a good process left unattended ceases to exist.

Chapter 16

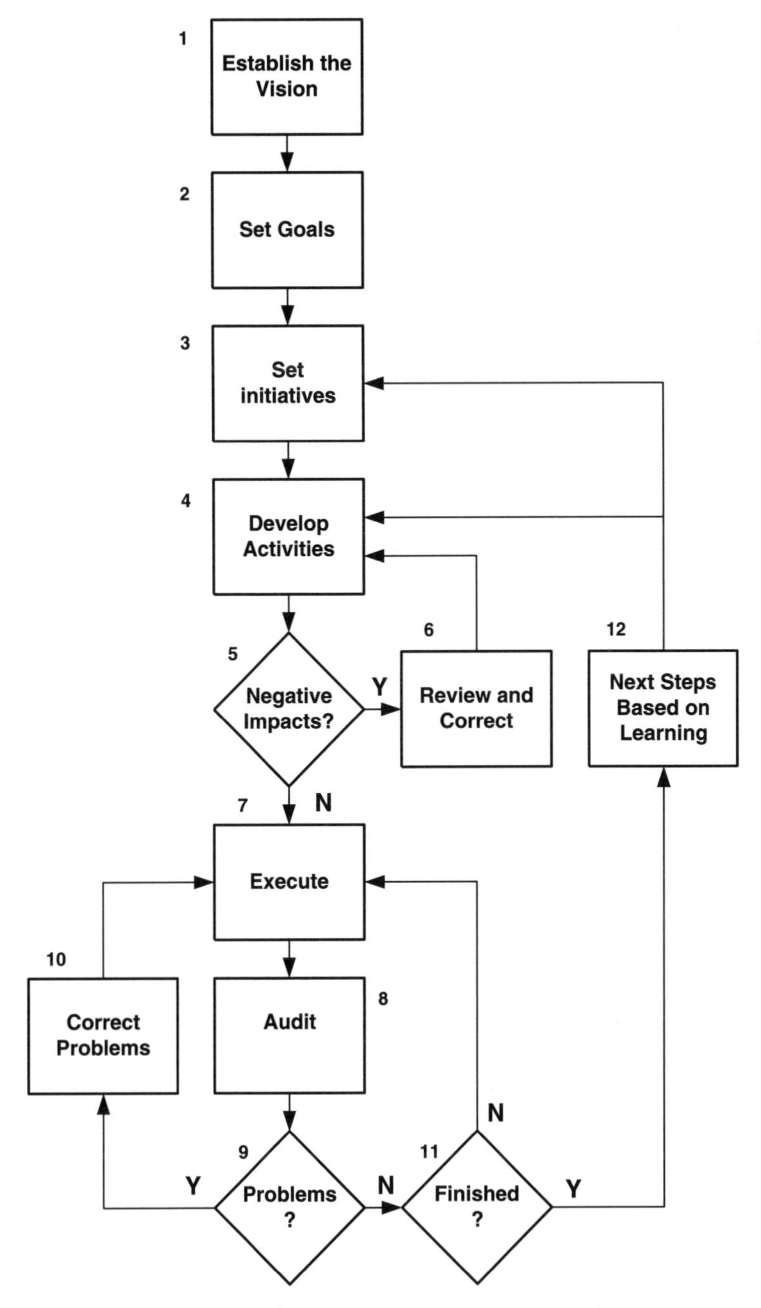

Figure 16-1 The Change Process Model

Chapter 16

16.2 The Change Process Model

In Chapter 11's discussion of work flows I provided recommendations about creating your own process diagrams. This section demonstrates the process in action, using the Change Process Model. This model, illustrated in Figure 16-1 describes how the process of change proceeds through implementation and beyond.

The diagram has twelve blocks, each representing a step in the process. The first seven are focused on getting the process started. The remaining five blocks represent the steps you need to continue the change process past implementation.

1. **Establish the Vision.** Block 1 marks where the process begins. This step was the subject of Chapters 4 and 5, and shapes the Goal Achievement Model.

2. **Set Goals.** As you will also remember from the Goal Achievement Model, goal setting is the next step. Goals are established to support the vision. The vision can be supported by more than one goal.

3. **Set Initiatives.** Use the goals to set initiatives that break the goals down into manageable pieces. These initiatives can be addressed by the work groups or teams.

4. **Develop Activities.** This step completes the Goal Achievement Model. At this point, specific activities should be developed. When they are completed, they will support the initiatives, goals, and vision established.

5. **Are There Negative Impacts?** Before implementing the change process, consider any possible negative impacts. Activities usually support the initiatives, goals, and ultimately the vision. Chapter 6 noted, however, the possibility that they might not. Block 5 gives you the opportunity to reflect on the impacts after you have developed the activities. This block marks a decision point. If you anticipate little if any negative impact, move on to Block 7. However, if you find that you have inadvertently created a negative impact, move to Block 6.

6. **Review and Correct.** You arrive at this block if you determine in Block 5 that your plan will create a negative impact. Now is the time to further review you Goal Achievement Model and make corrections to the activities. You then return to Block 4 because one or more of the activities will need to be adjusted. You then move back through Block 5 and ask the same question. If your answer still includes a negative impact, then you have not made the proper corrections. Return again to the loop made up of Blocks 4, 5 and 6. Eventually you will develop activities that do not negatively impact the process. You can then move on to Block 7.

7. **Execute**. This step is the one at which the change process is fully implemented. You execute the activities that you developed in Step 4. Because the process of change is never ending, you must evaluate how you are doing. The best way is audit the process. Auditing will be discussed in more detail in Section 16.3.

8. **Audit.** One way to check on the health of the effort is to audit the process. In this step, marked by Block 8, you make sure that you have mechanisms in place to identify process problems as the change is evolving. These mechanisms allow you to take corrective actions, to resolve any issues before they ruin what you and your team have taken so long to develop. Check the process frequently through auditing.

9. **Problems?** In this next block you analyze the results of the auditing process. In some cases, you won't find problems and can move onto Block 11. However, if problems are discovered, go to Block 10 to correct the problems identified by your audit.

10. **Correct Problems.** If you arrive at this block, you identified problems in the auditing of your change process. Some of these problems may be obvious early in the process. Some may not emerge until the process has been in place for a while. A single audit, therefore, may not be sufficient. In either case stay the course. Fix the problem. Put the corrective action back into Block 7's execution process. The loop consisting of Blocks 9, 10, 7, and 8 provides you the opportunity to take the corrective action needed for those problems associated with the change effort.

11. **Finished?** If your review of the audit indicates no problems, then you can move to Block 11. Now ask whether you have completed working on the activity. If you haven't, you don't necessarily have a problem, as indicated by your response in Block 9. Instead, you simply have not finished your work and you need to return to Block 7 to continue. At some point, the activity willbe completed and you will continue on to Block12.

12. **Next Steps Based on Learning.** This block builds on Chapter 3's discussion of spiral learning. Here you examine what you learned about your change effort. Use your analysis to drive a review and subsequent improvement in the initiatives (Block 3) and related activities (Block 4).

This model provides a good visual tool when you are trying to explain to others the continuous nature of change. Think what would happen if you stopped after Block 7, where you executed the original change process. Because events seldom, if ever, go exactly as planned, the first problem you encountered after execution would stop the process in its tracks. Do not allow this to happen. You have invested too much time and effort, and there is too much at stake for your organization.

16.3 Techniques for Maintaining Health

Auditing a process, represented by Block 8 in the Change Process Model, is nothing more than checking its health. The ultimate purpose is to find areas where problems already exist or, in a proactive environment, where they will exist. Once you find these problems, you then take appropriate action to solve them. The challenge is to audit without interrupting the change process, while at the same time getting sufficient detail to know when you need to correct problems before they get out of hand.

Before starting the auditing process, consider the following ground rules:

1. If you have problems, focus on their identification and resolution. Do not punish employees for not getting it right the first time. Instead, find out what went wrong and correct the process. Individuals almost never bear sole responsibility.

Usually it is the group, the organization, and sometimes even you may be the root cause of the problem.

2. If you are auditing the process for management, never reveal who gave you specific information. It is fine to provide a list of those who were audited, but do not break their trust by linking names to information. Good auditing is not a one-time event. You want people to trust you the next time around.

3. Do not create a bureaucratic nightmare in your effort to obtain information about the process. Others may believe that extremely detailed reports will provide the information needed to determine the health of what you have implemented. The bureaucratic nightmare is created when employees spend inordinate amounts of time writing reports instead of focusing on making the process work. Furthermore, many people want to avoid delivering bad news to the boss; therefore, the reports may not really tell you about problems.

4. Don't audit too soon or too late. If you audit the process too soon, you may not see the expected change. You may then react as if this lack of change was a problem, when in fact the change might just be starting to take effect. On the other hand, waiting too long can allow problems to become so large that they are difficult to correct. Some may not be correctable at all.

5. Self-auditing works at times. However, a third party who under stands what you are trying to do and who can be objective is often better. Self-auditing adds a bias to the process, one that can make things look better than they are.

With these ground rules in mind, we can now look at ways to audit the process that will not hamper the change process, but give you the information that you need to assess its health.

Third Party Assessment. In this method, someone who is familiar with your effort is brought in to assess its progress. This person could be someone from within the organization. If you have been working with consultants, they could also perform this task. Each of these has positive and negative aspects that you will need to weigh as you make your decision.

Internal resources are less expensive. However, unless they are from outside the department being assessed, they may have bias based on people who they know or have worked with in the past. External resources, such as consultants, cost more. At the same time, they can be far more objective. It's your choice; you just need to think it through before you decide which way to go.

You must guard against several pitfalls if you want the assessment to help the process. First, find someone who understands your process and will assess it. That person should not start suggesting changes to conform to his or her ideas. You do not need someone redesigning the process instead of trying to identify problems. Second, the number of interviews is critical. All too often, consultants who are hired to conduct interviews are given insufficient time or do not take the time they need. In either case, they interview too few people, then extrapolate their summary as if this small survey group was representative of the entire organization. In short, they make decisions about the health of the effort with too little information. Sometimes you can get away with this. More often, you arrive at the wrong analysis and cause undue harm to the organization. In one recent assessment that I conducted, I interviewed 50 people. The site thought that I was talking to too many. However, I was successfully able to assess the change effort and related work process, helping the site to identify improvement areas.

Be aware that a single assessment usually isn't sufficient, especially if problems are uncovered. The assessor should return after enough time has elapsed, to determine if the problems have been corrected. Ideally the same person should conduct the second set of interviews and do the assessment in order to maintain continuity through the process. In the assessment that I described above, I did have the opportunity to go back after a year to see if the issues that I had raised had been addressed. To the credit of those in the plant, the majority of the issues had not only been addressed, but the problems had been resolved.

Assessment by Walking Around. You can learn a great deal by simply walking around and talking to your people. Years ago, a book about managing by walking around described how you can learn vast amounts of information by simply going though the work area and asking employees how things were working. This is especially true when you are implementing a change initiative. If you listen, employees will tell you their problems and a whole lot more. You can then take this information, analyze it, and proceed through the problem correction process without a need for a full-blown work assessment. Be careful, however, to screen out the noise from the process problems. Although everyone will give you information, a lot of it may be flawed.

You gain other benefits from this type of effort. First, you have the ability to sell the change to the employees. You can not only solicit information, but also convince others that the change is for the better. Second, by talking to employees, you will be visibly demonstrating your support of the initiative.

Spot Audits. This approach is somewhat different from the full-blown assessment. In a spot audit, you check on a specific part of the process, keeping your review within these boundaries so that the audit has a very narrow scope. In addition, a spot audit is made on a random basis; employees don't know it is coming. Otherwise, they would temporarily clean up the specific problem and you would never see the process as it actually is under normal conditions. For example, compare your home when you know in advance visitors are coming and when you have no advance notice. In the former, you have a chance to clean up, presenting yourself in a way that may not match your normal living conditions. When people visit uninvited, they see you as you actually are. The same is true with the spot audit. To make spot audits most effective, they should be conducted by a team of auditors for the simple reason that the more people who are involved, the more depth there will be in the audit.

Measures. Measurement helps you determine the health of your change process. Select measures that are simple to understand, clearly measure what you want to look at when you assess the process, add value to the organization, and don't require a lot of effort to generate. The best measures point to what is going right or wrong. However, the wrong ones can cause a lot of problems for the organization and a lot of wasted work at the same time.

Oversight Team. One other way to check on the health of a change process is to put an oversight team in place. This team is usually formed by a group of senior managers who take ownership for the success of the effort. Their role is exactly what the name of their committee implies: to oversee the process, assess progress, and take corrective action when necessary. It is usually this group that authorizes the assessments and other methods of checking on initiative health. They also hold people accountable for correcting the problems and advancing the effort.

Because the oversight team is made up of senior managers, it serves as a role model for other employees. As the team stays involved through meetings and frequent monitoring, the employees see commitment that in turn will help them stay focused and succeed in their work.

16.4 Going Back is Never the Answer

Many in your organization will want to abandon the change initiative before it starts, during its implementation, or even after it has been put in place. Nevertheless, going back is never the answer. The older days may at times appear attractive, especially when you and your team struggle with the change process. Yet returning to past ways is not very good for many reasons:

- Change is a characteristic of a healthy organization. It may be an unsettling experience as the work and employee's jobs change. Yet change differentiates a company that will survive from one that won't.

- Whatever change is being implemented was not introduced on a whim. Change is usually the result of time and effort spent studying the environment, either external or internal, and then addressing issues or problems. Failure to change means failure to correct problems that will only get worse over time.

- Change is a constant in all businesses, even more in today's world than in past years. Change keeps companies competitive in a global market. If you fail to change, your competition will change without you. As a result, you may go out of business.

As individuals, we need to continually grow. A key part of that effort is changing how we work, how we use new tools to support our efforts, how we communicate, how we interact both internally and with our customers, and how we do most of our business processes in today's work environment. You may introduce new computers, new equipment, and new processes to your company, but there is one key factor still responsible for continued success. The people in your company need to grow and continuously improve. Otherwise, all other efforts and investments are wasted.

Chapter 17
Conclusion

17.1 Is This Really the End?

When you started reading this book, you most likely had one of several reasons:

- You were starting a change process and wanted information.
- You were involved in a change initiative and were looking for support.
- You had just finished one stage of a change effort and were starting the next.

Now that you have finished reading it, you have acquired new knowledge, new ideas, and alternate ways of thinking. In essence, you have just completed the kind of spiral discussed in Chapter 3. Although you are at the end of the book, you are still at the beginning of a much larger process that ultimately will help you positively influence change in your workplace.

17.2 Wrap Up

When I initially sat down to write this book, I wanted to present the material so that each chapter would build upon those that came before. Thus, as you progressed through the text, your store of information would grow, providing you with the tools you would need to work through the complex issues confronting you and your change initiative. We started with an introduction of terms you would need throughout the book. In Chapter 2, I presented the Ladder of Inference; making incorrect assumptions can get you in as much trouble as working with incorrect definitions. This was followed by the Learning Spiral, another important tool, especially for those of us who have been taught to think linearly. This new way of processing information gives you the freedom to learn as you grow. It also relieves you from the pressure of having to know the answer before you begin the work.

Next, we considered how to develop the overall vision. The Goal Achievement Model and the Roadmap provide the structural framework to support the vision. If they are followed, the organization will understand the reason for the change. It will also understand how each piece contributes and supports the overall plan.

Teams were then introduced, with special attention to the role they play in the work of change. The need to build teams is an easy concept to understand. Creating a team environment is a greater challenge. Teams can run counter to a culture that rewards individual effort. Yet in today's work environment, little if anything is achieved in a vacuum. When you review business successes, especially companies that avoided disaster by turning their world around, you find teams at the heart of the effort. The chapter about teams was logically followed by a discussion of consultants, who have the potential to be valuable team members. Some companies persist in viewing consultants as the enemy, yet they can be among the best resources supporting your change effort.

The discussion moved next to resistance. You can optimistically hope that everyone will follow your ideas and change initiatives without problems, but this view is probably incorrect. There will always be those who will resist. The chapter should help you not only recognize resistance (preferably before it takes hold), but also to overcome it.

With the stage now fully set, Chapter 10 introduced the Web of Change. This radar diagram looks at the elements of change in a new and different way. This discussion was continued in Chapters 11 and 12, where the individual elements were addressed in detail. Chapter 13 then addressed how to apply these chapters as well as the survey found in the appendix. Finally, the book discussed how to move the change process forward and how to track your effort with appropriate measurements.

Successfully Managing Change in Organizations was never intended to be a cookbook. You do not simply follow all the steps to achieve success. Change efforts don't work this way; they are not as orderly as a recipe. They are far more complex. Furthermore, even if I had wanted to create the perfect cookbook, I couldn't. That book would still be missing the single most important ingredient: you. Everything that I have shared with you in this book now needs to be considered from your perspective and then applied to your work place for your change initiative.

17.3 The Book Accompanies You on the Journey

This book does more than start you on the road; it also accompanies you on the journey. My own challenge is how to address my goal of staying in touch with my readers? The usual book is a one-way street. You buy it, and you read it. Unlike the communication model introduced in Chapter 12, the usual book typically provides no feedback. This lack of feedback is a problem that I want to address.

As you have worked your way through the text, many things have happened.

- You have learned a lot of new information.
- You have related the content to examples of your own.
- You have developed questions.
- You have had experiences and new ideas that you want to share.
- You have had new ideas for the web survey.
- You have wondered what others were doing.

The first item in this list fits perfectly into the conventional book format, with one-way communication from me to you. The other items ask for two-way communication. Fortunately technology—in this case, the Internet—provides the answer.

I have established a web site that you will be able to reach, through a link from the Industrial Press web site or directly, once you have my address. You will be able to use my web site to

- Ask questions about the tools introduced in the book
- Share stories about your successes, failures, problems and problem resolutions
- Send your web survey scores and any suggestions for improving the survey

Through the web site, we will share a two-way dialog. I am also planning to set up the web site in a way that I can share some of your stories and ideas with other readers who may be having similar problems. Sources will be kept confidential. To find the web site, go to www.industrialpress.com and search by my name to locate the link.

Chapter 17

Finally, let me conclude with a quotation (source unknown) that summarizes the effort of change management in organizations.

When you reach the end of the known and are ready
to step out into the dark, you need to have courage
and believe that you will find firm ground

Change in our work is that step into the unknown. I assure you, though, that with dedication and commitment to your vision, you will find firm ground.

Goodluck!

Steve Thomas

References

Agryis, Chris. *Overcoming organizational defenses: facilitating organizational learning.* New Jersey: Prentice-Hall, 1990

Barstow, Alan. "Building Effective Organizations: Getting Things Done." Masters course presented at the University of Pennsylvania. Philadelphia, PA., 1995.

Bartol, Kathryn M. and David C. Martin. *Management.* New York: McGraw-Hill, 1998.

Eldred, John. "Mastering Organizational Politics and Power." Masters course presented at the University of Pennsylvania. Philadelphia, PA., 1994.

Katzenbach, Jon R. and Douglas K. Smith. *The wisdom of teams.* New York: HarperCollins Publishers, 1993.

Kayser, Thomas A. *Mining group gold.* California: Serif Publishing, 1990.

Senge, Peter et al. *The Fifth discipline fieldbook: strategies and tools for building a learning organization.* New York: Doubleday Press, 1994

Thomas, Stephen J. "Creation of an Area Work Team Environment at a Plant Site." Master's thesis, University of Pennsylvania, 1995.

Wing, Kennard T. "Why most organization change efforts fail," u.m., 1993

Web of Change Survey

Introduction

The Web of Change survey that follows is described in Chapter 10 through 13. You can choose two ways to proceed. First you can create your own web diagram, as described in Chapter 10. To make this easier, I have included a blank web diagram with the survey. Using this diagram, you can conduct the survey that follows, total the scores for each section (Figure A-1), and plot them on your web (Figure A-2). I have also provided an alternative. At the back of the book is a disk that uses Microsoft EXCEL97® to automate the process. The disk has a worksheet tab for each of the eight elements as well as summary pages for the baseline and the reassessment. These tabs appear at the bottom of the screen. The software also validates your score so that you can not give yourself more than the allowable points per question. The charts are automatically created as you score the individual questions. Both methods work. Choose the one that best fits your circumstances.

Directions:

The directions for completing both the manual and computerized survey are as follows:

Web Survey

Baseline

1. Complete all questions in each of the eight sections of the survey. If you are not completely sure of the answer, select the one that most closely matches what you believe is the answer. All questions must be answered for the survey and scoring to be complete.

2. Each section has five questions. Each question can be scored a maximum of 4 points. Therefore, the maximum score for any section is 20. All questions are of the same type.

3. Total the scores for each section as you finish. If you are conducting the survey manually, enter the total on the summary sheet (Figure A-1), then onto the web diagram (Figure A-2). If you are using the Microsoft EXCEL spreadsheet, the totals will be entered automatically.

Reassessment

The directions for conducting the reassessment are the same as those for the baseline. The only difference is that, on the Microsoft EXCEL spreadsheet, you use the reassessment column.

Web Survey

CATEGORY	TOTAL CATEGORY SCORE
A. Leadership	
B. Work Process	
C. Structure	
D. Group Learning	
E. Technology	
F. Communication	
G. Interrelationships	
H. Rewards	

Figure A-1

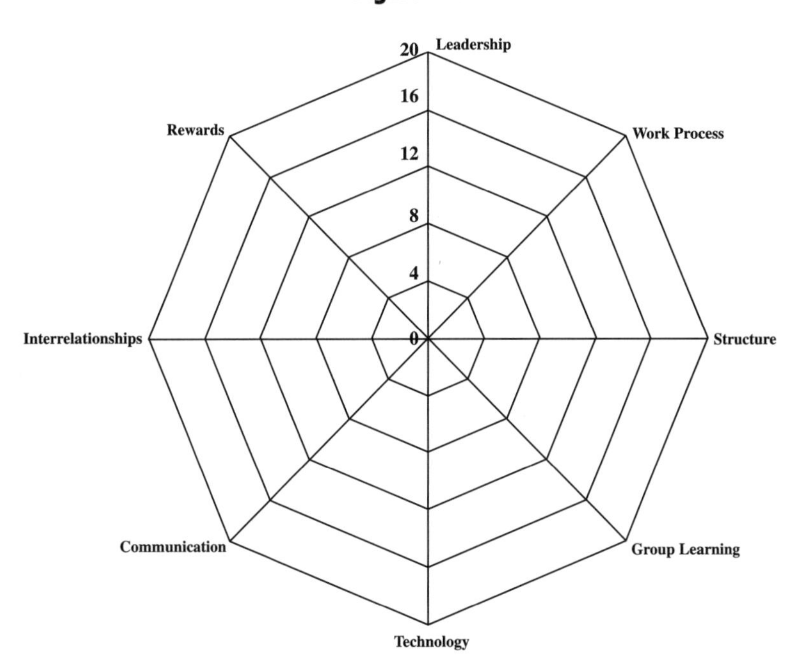

Figure A-2 Basic Web Diagram

Web Survey

Survey of the Elements

A. Leadership

Points

1 Rate the reaction of the site personnel to the leadership of the management team.

A. High level of effort to follow leaders	4
B. Average level of effort to follow leaders	3
C. Minimal level of effort	2
D. The organization does not support the leaders.	1
E. There is outright resistance to the leadership.	0

2 The leadership is transitional (leading the organization towards a vision of the future). They are not transactional (concerned only with activities of the day).

A. Strongly agree	4
B. Agree	3
C. Neutral	2
D. Disagree	1
E. Strongly disagree	0

3 Management allows the organization freedom to act within a set of predefined boundaries without micromanaging of its efforts.

A. Strongly agree	4
B. Agree	3
C. Neutral	2
D. Disagree	1
E. Strongly disagree	0

4 When the organization develops and implements goals, initiatives, and activities, - those involved are held accountable for success.

A. Strongly agree	4
B. Agree	3
C. Neutral	2
D. Disagree	1
E. Strongly disagree	0

5 The leaders in the organization remain in their positions long enough to live with the results of the changes they initiate.

A. Strongly agree	4
B. Agree	3
C. Neutral	2
D. Disagree	1
E. Strongly disagree	0

Web Survey

B. Work Process	**Points**
1 Work processes have been developed for conducting work.	
A. Strongly agree	4
B. Agree	3
C. Neutral	2
D. Disagree	1
E. Strongly disagree	0
2 Work processes are reviewed and updated as improvements are made or as new and improved ways of doing business are learned.	
A. Strongly agree	4
B. Agree	3
C. Neutral	2
D. Disagree	1
E. Strongly disagree	0
3 Measurements are in place to determine if the processes are working properly. Corrective action is taken if they are not.	
A. Strongly agree	4
B. Agree	3
C. Neutral	2
D. Disagree	1
E. Strongly disagree	0
4 The processes were designed to represent a desired future state so that the site can use them as goals for achieving excellence.	
A. Strongly agree	4
B. Agree	3
C. Neutral	2
D. Disagree	1
E. Strongly disagree	0
5 The processes were developed by a team of experts (internal or external) whose goal was to improve how work is accomplished.	
A. Strongly agree	4
B. Agree	3
C. Neutral	2
D. Disagree	1
E. Strongly disagree	0

C. Structure

Points

1 The organization's structure was designed and implemented to support the business strategy.

A. Strongly agree	4
B. Agree	3
C. Neutral	2
D. Disagree	1
E. Strongly disagree	0

2 The structure supports the work processes within a department and between departments (across the departmental interfaces).

A. Strongly agree	4
B. Agree	3
C. Neutral	2
D. Disagree	1
E. Strongly disagree	0

3 The structure supports people working together in teams to accomplish the business objectives.

A. Strongly agree	4
B. Agree	3
C. Neutral	2
D. Disagree	1
E. Strongly disagree	0

4 The structure was designed to support delegation of authority and communication both internally (within a department) and externally (between departments).

A. Strongly agree	4
B. Agree	3
C. Neutral	2
D. Disagree	1
E. Strongly disagree	0

5 The levels in the structure from the bottom of the organiza-tion (the work force) to the top (the manager of the plant or facility) appears to be correct in that they support effective execution of the work.

A. Strongly agree	4
B. Agree	3
C. Neutral	2
D. Disagree	1
E. Strongly disagree	0

Web Survey

D. Group Learning — Points

		Points
1	**Rate the level of management support for learning new, different and better ways of working within the organization.**	
	A. Very supportive	4
	B. Supportive	3
	C. Neutral	2
	D. Non supportive	1
	E. It is not present at the site	0
2	**The personnel in the organization are allotted time for learning and training each year (excluding mandatory training) and are encouraged to use the time.**	
	A. Strongly agree	4
	B. Agree	3
	C. Neutral (neither agree or disagree)	2
	D. Disagree	1
	E. Strongly disagree	0
3	**Training and group learning are not aimed just at job performance skills, but also at skills that promote better interaction and team work.**	
	A. Strongly agree	4
	B. Agree	3
	C. Neutral	2
	D. Disagree	1
	E. Strongly disagree	0
4	**When training is to be developed, an effort is made to determine the difference between what employees already know and what they need to know.**	
	A. Strongly agree	4
	B. Agree	3
	C. Neutral	2
	D. Disagree	1
	E. Strongly disagree	0
5	**Employees have input into what they are going to learn.**	
	A. Strongly agree	4
	B. Agree	3
	C. Neutral	2
	D. Disagree	1
	E. Strongly disagree	0

Web Survey

E. Technology

Points

1 Software tools which support the work process exist and are well utilized.

A. Strongly agree	4
B. Agree	3
C. Neutral	2
D. Disagree	1
E. Strongly disagree	0

2 The software is integrated into a single electronic system in order to simplify use and avoid multiple entries and databases that could cause problems.

A. Strongly agree	4
B. Agree	3
C. Neutral	2
D. Disagree	1
E. Strongly disagree	0

3 The data within the system is easily accessible to all employees who need it.

A. Strongly agree	4
B. Agree	3
C. Neutral	2
D. Disagree	1
E. Strongly disagree	0

4 The plant has a strategic multiyear plan for enhancing software in order to stay current with the technology and to provide high levels of support to the workforce.

A. Strongly agree	4
B. Agree	3
C. Neutral	2
D. Disagree	1
E. Strongly disagree	0

5 Support that is readily available for both the users and the system exists at the site.

A. Strongly agree	4
B. Agree	3
C. Neutral	2
D. Disagree	1
E. Strongly disagree	0

Web Survey

F. Communication

		Points
1	**The leaders of the organization (at the site level) believe that communication is important and act accordingly**	
	A. Strongly agree	4
	B. Agree	3
	C. Neutral	2
	D. Disagree	1
	E. Strongly disagree	0
2	**Events that occur at the site are well communicated so that employees are aware of what is going on.**	
	A. Strongly agree	4
	B. Agree	3
	C. Neutral	2
	D. Disagree	1
	E. Strongly disagree	0
3	**Meetings include: an agenda, a preestablished start and stop time, facilitation, and minutes, with action items, of what was discussed.**	
	A. Strongly agree	4
	B. Agree	3
	C. Neutral	2
	D. Disagree	1
	E. Strongly disagree	0
4	**Communication tools (intranet, voice mail, e-mail, phones and radios) are provided to those who need them .**	
	A. Strongly agree	4
	B. Agree	3
	C. Neutral	2
	D. Disagree	1
	E. Strongly disagree	0
5	**The communication systems identified in item #4 are effective, providing the needed level of communication among site personnel.**	
	A. Strongly agree	4
	B. Agree	3
	C. Neutral (neither agree or disagree)	2
	D. Disagree	1
	E. Strongly disagree	0

G. Interrelationships — Points

1 The vertical working relationship across the organization is developed to a level that supports change initiatives.

A. Strongly agree	4
B. Agree	3
C. Neutral	2
D. Disagree	1
E. Strongly disagree	0

2 The horizontal working relationship across departments is developed to a level that supports change initiatives.

A. Strongly agree	4
B. Agree	3
C. Neutral	2
D. Disagree	1
E. Strongly disagree	0

3 When teams are formed to work on projects (including implementation of change), the relationships are well developed and mutually supportive.

A. Strongly agree	4
B. Agree	3
C. Neutral	2
D. Disagree	1
E. Strongly disagree	0

4 The working relationship between the workforce (union or nonunion) and management is supportive of change initiatives.

A. Strongly agree	4
B. Agree	3
C. Neutral	2
D. Disagree	1
E. Strongly disagree	0

5 The working relationship between the various locations (e.g., corporate head quarters and plant sites) is mature to the point that best work practices are shared and used.

A. Strongly agree	4
B. Agree	3
C. Neutral	2
D. Disagree	1
E. Strongly disagree	0

Web Survey

H. Rewards

Points

1 The financial success of the business is shared through a gain or profit sharing plan with those who contributed.

A. Strongly agree	4
B. Agree	3
C. Neutral	2
D. Disagree	1
E. Strongly disagree	0

2 A reward structure exists that links financial reward to the success of teams and work group (as opposed to the individual).

A. Strongly agree	4
B. Agree	3
C. Neutral	2
D. Disagree	1
E. Strongly disagree	0

3 An individual merit plan exists, but much of the input comes from managers, subordinates, and peers evaluating how the individual performed as a member of the team.

A. Strongly agree	4
B. Agree	3
C. Neutral	2
D. Disagree	1
E. Strongly disagree	0

4 There is a viable process in place to provide short-term rewards to reinforce change efforts by the organization.

A. Strongly agree	4
B. Agree	3
C. Neutral	2
D. Disagree	1
E. Strongly disagree	0

5 Rate the reward system based on your perception of its ability to create rewards based on individual and team performance.

A. Excellent	4
B. Good	3
C. Average	2
D. Poor	1
E. Very Poor	0

INDEX

S

T

U-V

W-X-Y-Z